Participatory Politics

M000205371

This report was made possible by the grants from the John D. and Catherine T. MacArthur Foundation in connection with its grant-making initiative on Digital Media and Learning. For more information on the initiative, visit http://www.macfound.org.

The John D. and Catherine T. MacArthur Foundation Reports on Digital Media and Learning

A complete series list can be found at the back of the book.

Participatory Politics

Next-Generation Tactics to Remake Public Spheres

Elisabeth Soep

The MIT Press
Cambridge, Massachusetts
London, England

MIT Press books may be purchased at special quantity discounts for business or sales promotional use. For information, please email special_sales@mitpress.mit.edu.

This book was set in Stone Serif and Stone Sans by the MIT Press. Printed and bound in the United States of America.

Soep, Elisabeth.
Participatory politics : next-generation tactics to remake public spheres / by Elisabeth Soep.
    pages   cm. — (The John D. and Catherine T. MacArthur Foundation reports on digital media and learning)
Includes bibliographical references and index.
ISBN 978-0-262-52577-0 (pbk. : alk. paper)
1. Youth—Political activity. 2. Political participation—Technological innovations. 3. Internet and youth—Political aspects. 4. Community leadership. I. Title.
HQ799.2.P6S64   2014
320.40835—dc23
2013028481

10 9 8 7 6 5 4 3 2 1

Contents

Series Foreword

The John D. and Catherine T. MacArthur Foundation Reports on Digital Media and Learning, published by the MIT Press in collaboration with the Monterey Institute for Technology and Education (MITE), present findings from current research on how young people learn, play, socialize, and participate in civic life. The reports result from research projects funded by the MacArthur Foundation as part of its $50 million initiative in digital media and learning. They are published openly online (as well as in print) in order to support broad dissemination and to stimulate further research in the field.

Acknowledgments

I am so grateful to be part of the following network of researchers who codeveloped the concept of participatory politics offered here, produced key studies, and shared crucial feedback as I prepared this piece: Danielle Allen, Cathy Cohen, Jennifer Earl, Elyse Eidman-Aadahl, Howard Gardner, Mimi Ito, Henry Jenkins, Joseph Kahne, and Ethan Zuckerman.

The work could not have happened without the exceptional vision and support of Connie Yowell and An-Me Chung from the John D. and Catherine T. MacArthur Foundation's Digital Media and Learning Initiative.

Sangita Shresthova, Ellen Middaugh, and Carrie James—all leaders of Youth and Participatory Politics study teams—have been enormously generous and thoughtful in guiding me through their projects and sharing the most striking early findings they and their colleagues have discovered in the last few years of concerted research.

Ellen Seiter, the editor of this MIT reports series, has read and reread the piece at various stages and seriously pushed my thinking—thank you! Pendarvis Harshaw, Rebecca Martin, and Shirin Vossoughi shared reflections that have been instrumental in my efforts to get it right.

Of course, I take full responsibility for all that is here.

Introduction

In 2012, 24-year-old Pendarvis Harshaw was finishing up some college courses and working as a mentor for the local school district. Over spring break, he set off on a road trip to visit his father, whom he hadn't seen in 18 years. It was through his uncle on Facebook that Pen had tracked down his dad. Pen flew from Oakland to Chicago and then joined a friend with a car for the 12-hour drive to the Alabama prison where his father was incarcerated. Pen tweeted the whole way, regularly updating his growing community of 2,250-plus followers.

A couple of months later, he wrote a story about the experience. An emerging journalist who had spent his teen years at Youth Radio, a youth-driven production company in Oakland, Pen set out to get the story distributed. The week before Father's Day, he published the story online, hoping a big outlet would pick it up. There were no takers. So when Father's Day arrived, he posted the piece on his own Tumblr, *OG Told Me*, a photo-rich oral history site chronicling his encounters with black male elders and their advice to young men.

Soon after, Pen wrote on Facebook, "After pitching my piece about my journey to meet my father to multiple outlets that report on 'Black news'—and getting no response . . . I decided to post it on my personal blog. In turn, the response from my circle of friends has been amazing." Pen's friends had commented on the article and spread the link, urging others to do the same. They reflected on how the story touched them personally and connected it to issues such as mass incarceration, drug policies, the role of journalism in public affairs, race and masculinity, and fatherhood.

What really got to Pen, though, were the in-person responses. When he was jogging around Lake Merritt one day, an acquaintance stopped him to say, "I didn't want to react online, but I wanted to tell you in person how much I appreciated your being so transparent and open." On Facebook, Pen wrote, "Conversations, texts, emails, tweets, facebook shares & likes . . . all from a lil sumn I decided to write . . . that's love. Thanks."

Pen did not overturn a government, get an official hired or fired, or change a policy by producing, sharing, and stoking conversation through his story about his father. He did, however, engage in some of the key activities that drive youth involvement in civic life today. This emerging set of activities fuels what my colleagues and I are calling *participatory politics*.

In the United States, around 2012, there were several epic examples of participatory politics, in which young people used digital and social media to exercise voice and agency on issues of public concern (Cohen, Kahne, Bowyer, Middaugh, and Rogowski 2012; see also Kahne, Middaugh, and Allen, forthcoming). The Occupy movement live-streamed massive public demonstrations against economic inequality from encampments across the country. A controversial call to action originating in Southern California against a Ugandan war criminal became the most popular upload in YouTube's history. Young people lit up their social networks to express opposition to federal legislation they believed would limit Internet freedom.

I will discuss these events and many others in the course of this report. Pen's example offers a less sensational but equally revealing case of the everyday ways in which young people are merging the cultural and the political to understand, express, and reshape public affairs.

Though sometimes disavowing "politics" as an apt description of what they're doing, civically engaged young people are using every means and medium at their disposal to carry out the core tasks of citizenship. Through a mix of face-to-face and digital encounters and interventions, they deliberate on key issues, debate with peers and powerbrokers, and in some cases change the structures of joint decision making and the course of history (Allen and Light, forthcoming). Like Pen, many young people who are coming into their political selves today both distrust public institutions and want in. They get excited about alternative ways to make a difference, and they seek access to traditional channels to power. They may appear to act alone but are always operating in interconnected networks that allow for and inhibit specific modes of civic engagement. Through their interactions with peers and elites, they are redefining some key dynamics that govern civic life.

This report delves into these shifting dynamics to ask the following: What specific tactics are young people experimenting with to exercise agency and intervene in public affairs? How can these activities grow in quality? What work is required to ensure that opportunities to engage in participatory politics are equitably distributed among youth, including youth who are marginalized from digital access and other forms of privilege? I will draw from insights in the existing literature as well as in a set of interconnected though independent studies that are still ongoing, as part of the Youth and Participatory Politics (YPP) Research Network, an initiative supported by the John D. and Catherine T. MacArthur Foundation. YPP research includes the following:

• Ethan Zuckerman's collection of case studies designed to explore the dynamics of activism in the age of digital communications.

• Danielle Allen's cross-site, interdisciplinary series of studies of the public spheres of the contemporary United States and other nations and the role of the Internet in them.

• An interview-based national study from Howard Gardner, Carrie James, and colleagues centered on young people's civic and political participation in the digital age.

• A set of qualitative case studies of exemplary youth organizations and networks that encourage productive forms of participation in public spheres, from a team headed up by Henry Jenkins and Mimi Ito.

• Cathy Cohen and Joe Kahne's national survey of young people tracking the quantity, quality, and equality of their new media practices and civic attitudes and behaviors.

• Jennifer Earl's multimethod examination of youth-related protest from both the youths' and the targets' perspectives.

• Elyse Eidman-Aadahl's work engaging education practitioners, inside and outside school, to theorize new practices in support of youth civic engagement and participation in networked public spheres.

• My own participatory research investigating the production and digital afterlife of youth-made media and mobile technology development aimed to advance the public good.

My participation in this network of researchers bridges the role of scholar and practitioner. Since 1991 I have researched youth learning and civic engagement in community organizations and peer-based activities in which young people create media projects aimed to engage, inform, and move the public. Starting in 2000, I added a second set of responsibilities to that effort when I began actively collaborating with youth and adult colleagues as a hands-on media producer and educator. Together we have cocreated radio stories, videos, photo essays, online posts, spoken word poetry projects, and mobile apps, sometimes reaching audiences in the tens of millions, all the while documenting our joint process using ethnographic and participatory research techniques (see, e.g., Soep 2005, 2012; Soep and Chávez 2010).

The frameworks emerging from the YPP Research Network are designed to be more than abstract concepts; they must advance the work at hand. Alongside other action-oriented colleagues, I am always asking: How effectively do these frameworks capture the range of activities young people carry out on a day-to-day basis as they develop civic practices, products, and theories of change? How useful are these frameworks for educators, producers, advocates, and organizers working in diverse online and offline settings to support young people's agency in public spheres?

The list of tactics identified here is meant to spark discussion and debate about what is covered, what is missing, and what further work is required to understand and support the highest quality and most equitably distributed forms of participatory politics. The tactics are derived from a systematic research agenda and from sustained, direct collaboration with youth. My aim is for these tactics to resonate with and advance the efforts of young people and their allies who are doing the work of participatory politics on shifting ground, where the crucial matter of young people's role in democracy is in question and at stake.

The five tactics emerging out of this combination of research and practice are as follows:

1. Pivot your public
2. Create content worlds
3. Forage for information
4. Code up
5. Hide and seek

After reading a detailed explanation of these tactics as they manifest in a sweeping range of youth-driven activities across the United States, you will find a discussion of concrete ideas for cultivating the new literacies we'd better invest in if we want young people in various life circumstances to have a voice in and a shot at shaping public affairs. If the tactics suggest *how* civically engaged young people are exercising agency in public life, literacies tell us something important about the *know-how* they rely on to do that work effectively. Participatory politics don't come automatically, even for young people raised on mobile devices and digital media. Nor do individuals act alone when they deliberate and pursue justice, and in this sense it's best to frame literacies as activities that communities can organize themselves around through interconnected efforts, rather than as skills possessed by or lacking in this or that young person or segment of the youth population.

In Pendarvis's case, it took a strategy for him to maintain robust and receptive networks of peers over time so they were there for him when he needed them, and he relied on platforms that allow for disparate communities to act, when necessary, in concert. For Pen to understand how his story was spreading through his blog and social networking sites, he consulted with a geek who lived one floor up in his apartment building who taught him how to run the numbers on various analytics programs. Pen regularly boosted friends' projects rather than only promoting his own, and he updated frequently and then periodically dropped off for 72 hours so people would notice his absence and say so in tweets like "Where's Pen? Are you lost in Oakland?" Knowing how to monitor and interpret complex data sets, including those generated by one's own moment-to-moment activities, and knowing how to stoke the engagement of peers without being too obvious about it are, it turns out, key literacies associated with civic engagement in the digital age.

After a discussion of literacies is a section on risk. While opening new opportunities, the tactics associated with participatory politics also raise a series of concerns and critiques that merit serious attention, including simplification, sensationalization, slippage, unsustainability, and saviorism.

Finally, the paper addresses shifting dynamics that underlie youth-driven politics in the age of digital communications, and it concludes with implications for future research and on-the-ground activity.

Participatory Politics: What Sets It Apart?

Consider some of the activities Pen and his community carried out in his storytelling project. They *circulated information*, activating various channels, including self-publication through personal outlets, while also pursuing third-party distribution. They *sparked dialogue*, not only telling but also hearing; Pen deliberately stoked conversation by joining in comment streams, publicly recognizing "link-love" when others reposted his piece, and by warmly receiving acknowledgment of what he'd shared in his story. Both he and his readers *produced content*, using digital tools and platforms to craft narratives and responses that addressed potent social themes. They *investigated* sources of information, connection, and opportunity, not only in the service of crafting the story but also to discover and track channels for reaching and activating significant audience. Finally, they *prodded others to act*. Peers confessed that they were motivated to reach out to members of their own families and continue their work on issues relevant to Pen's story, based on his words.

These five activities that powered Pen's project are the core features of participatory politics (Kahne, Middaugh, and Allen, forthcoming):

Circulation In participatory politics, the flow of information is shaped by many in the broader community rather than by a small group of elites. Circulation might include sharing information about an issue at a meeting of a religious or community organization to which one belongs or posting or forwarding links or content that have civic or political intent or impact.

Dialogue and feedback   There is a high degree of dialogue among community members, as well as a practice of weighing in on issues of public concern and on the decisions of civic and political leaders. This might include commenting on blogs or engaging in other digital or face-to-face efforts to interact with or provide feedback to leaders.

Production   Members not only circulate information but also create original content (such as a blog or video that has political intent or impact) that allows them to advance their perspectives.

Investigation   Members of a community actively pursue information about issues of public concern. Rather than simply relying on established, elite-driven sources of information, participants seek out, collect, and analyze information from a wide array of sources. They also often check the veracity of information that is circulated by elite institutions, such as newspapers and political candidates.

Mobilization   Members of a community rally others—ranging from diffuse friendship groups and online networks to organized groups focused on related issues—to help accomplish civic or political goals. This might include working to recruit others for a grassroots effort within one's community or reaching out to those in one's social network and beyond on behalf of a political cause.

Young people have always circulated media in various forms, engaged in dialogue, produced content, investigated their worlds for information and insights, and mobilized peers toward shared goals. But evidence suggests that these activities have become less centralized and more prominent in the context of civics, in large part because of the dynamics of digital and social media. Participatory politics build on and reinforce three important shifts operating at the level of the individual, the collective, the institution, and the systems that connect all three.

First, the gap between culture and politics is shrinking. That much is evident in the extent to which the core features of participatory politics mirror the same activities we see in highly active participatory cultures. For decades, Henry Jenkins and colleagues have studied the communities energized by popular-culture products, including television shows, games, comic book series, and film and book franchises. This research has shown how young people move far beyond the position of passive consumer and emerge as highly engaged, imaginative, active coproducers of media and community. When governed by low barriers to involvement, a mandate to connect and share ideas and creations with peers, and informal mentorship systems that enable newcomers easily to learn from veterans, popular-culture worlds can convert audiences into extremely dedicated and productive makers of meaning and media (Jenkins 2008). Likewise, when these same conditions are in place around matters of civic import, young people utilize the vocabulary, symbols, communities, and rituals of popular culture to voice their opinions and exert influence on matters of public concern.

Second, young people turn more and more toward peers to carry out the work of civics. They use social media sites and mobile devices to circumvent gatekeepers, reduce costs by diminishing the importance of physical copresence, facilitate states of constant connection, and amplify the outcomes of those connections for widening circles of friends and followers (Earl and Kimport 2011). The issue is not that "the people formally known as political elites" cease to matter as targets and allies in civic activity; rather, there are new ways to reach those elites—for example, online petitions and Twitter campaigns.

Nascent elites are also forming. Young people who can bang out the right string of code just in time, who have networks ready to act, or who know how to create engaging transmedia stories with irresistible calls to action have joined the ranks of key civic influencers. Traditional measures such as voting rates, social studies test scores, and counts of youth participants at street protests remain important indicators of civic knowledge and engagement. But participatory politics calls for new strategies attuned to network effects and discourse flows to assess youth involvement in shaping public spheres.

Third, it is easy to get excited about the new openness that can be facilitated by less hierarchical structures for communication, but we also need to watch the new inequalities that can block access to the knowledge and networks that drive today's change. Freedom, after all, is not possible without equity, argued Danielle Allen (2012). Politics are a kind of art, she said, and participation requires the mastery of techniques through which citizens can understand shared experience, see and pursue alternative paths, "take turns accepting losses in the public sphere, and . . . acknowledge and honor the losses that others have accepted" (p. 1). Interdependence along these lines is required to produce democracy; hence the focus in participatory politics on collective practices across a range of tools and platforms that aim to promote freedom, equity, and democratic deliberation.

A final point centers on a small but crucial word in the prior sentence: *aim*. How much does aim, or *intention*, matter in participatory politics? On the one hand, is it legitimate to frame an activity as political if the young people involved are mainly in it for fun, or if they can't or don't choose to articulate a thoughtful political logic behind their activities? On the other hand, do we assign meaning to young people's activities in the civic realm only to the extent that they yield observable *effects*? What if young people learn a great deal by mounting a politically motivated action among peers that utterly fails? What if, like Pen in the opening example, young people compose and share stirring personal narratives touching on issues of public concern, and doing so builds the capacity for deepened civic participation within themselves and their communities further down the line—yet no concrete, policy-oriented mobilizations take place in the moment? Should we still consider that initial work as evidence of participatory politics?

In answering questions such as these, we must realize that context plays a hugely important role in shaping the meanings, motives, and effects of young people's political engagements (Bourdieu 1977; Chaiklin and Lave 1996; Ortner 1984). Rather than judge the value of young people's civic participation based on the individuals' intentions, on the one hand, or their activity's impact, on the other hand, we need to track the political becoming of youth and the direct and indirect effects of their social practices in public spheres over time. Rather than superimpose generic measures of political potency, we need to grapple with the explicit and tacit "theories of change" young people and their collaborators pursue through their civic activities— whether, for example, they seek to transform policy, sway elites, render new services, or reframe issues and identities at the level of culture (Zuckerman 2012a).

Tempting as it is, the goal should not be to determine once and for all whether any new tech platform or tool promotes freedom or oppression; that line of thinking overassigns power to technology and underaccounts for the mixing of emancipation and exclusion in just about any civic undertaking worth taking seriously. So we need ways to assess the likelihood that any of the tools, activities and theories of young people will bring about the desired effects.

"A lot of things can influence the political process that you might not have considered political until it's had that influence," said one subject in a YPP study of libertarian youth in the United States. "Who would think that a hurricane would be political?" she pointed out, speaking of Katrina, "until it was?" (Thompson 2012).

Five Tactics of Participatory Politics

1. PIVOT YOUR PUBLIC

Mobilizing civic capacity within networks that form out of shared personal and popular culture interests and communities.

The opening example of Pendarvis Harshaw's distribution strategy for a story ignored by mainstream news outlets is an instance of pivoting his public. As noted earlier, Pen's friends and followers shared many personal, social, and cultural interests. That much is obvious from his social media posts about upcoming poetry events and bicyclist gatherings and his photos of extreme hairstyles. By including news of his journey in his ongoing social media updates—where he also cheered the Oakland As and planned his upcoming birthday celebration—Pen enlisted a network of already connected friends and associates to examine issues relevant to public affairs. Personal and playful updates were interspersed with links to Pen's own critical writing, to conversation about President Barack Obama's responsibilities as a mentor to the next generation, and to provocations such as "Confused. So . . . They're closing public schools across the Nation & privatizing education? What does it mean?"

He converted these online activities into meaningful offline interactions, defying the troubling extent to which observers tend to bifurcate young people's digital and real-life activities (and to argue about whether e-politics "count" without acknowledging the constant crisscrossing of virtual and face-to-face encounters in most civic and political activities). What's at work here is the tactic of marshaling the shared history and sensibility that can form inside a network of mutual interest and using this to motivate engagement in topics and activities with civic import.

Civic potential within these networks may appear to be latent, but as Pen's strategies and wider YPP research suggests, a great deal of work goes into maintaining communities so they are ready to mobilize on behalf of collective efforts, under the right conditions (Kligler-Vilenchik, forthcoming). Howard (2010, p. 12) has argued that "banal tools for wasting time"—he named Facebook and YouTube—serve as an infrastructure for social movements during times of political crisis. Participatory politics highlights the civically grounded uses of these tools, even in the easiest of times. Henry Jenkins and others have convincingly demonstrated the significant creative work and literacy development that happen when peers pursue their popular-culture interests together, and the value of these activities should not be judged exclusively on the basis of whether they translate directly into civic outcomes (Ito 2009; Jenkins 2008; Jenkins, Purushotma, Clinton, Weigel, and Robison 2009). Moreover, we have seen that through their interest-driven activities, young people are often crafting, feeding, and nurturing a community that will be prepared to think and act in political ways when the appropriate time comes.

Survey data from a national study of youth and participatory politics provides further evidence of how interest-driven activities can power civic engagement, perhaps especially in the digital age. It appears that such activities lay a foundation for participatory politics by cultivating "digital social capital" (Cohen et al. 2012). Pursuing interests can build knowledge, skills, and networks that support subsequent (or simultaneous) political organizing. Cohen and colleagues found that young people who were highly involved in interest-driven activities were five times as likely as those without such involvements to engage in participatory politics, and they were nearly four times as likely to participate in all political acts measured in the survey.

A longitudinal study centered on students in Chicago's Digital Youth Network reported a related finding. Students in that program who were classified as "highly engaged" were ten times as likely as their less engaged peers to imagine using media to effect social change in the future (Barron, Gomez, Pinkard, and Martin, forthcoming). Engagement levels were based on the depth of digital media production experience students pursued during the three years of classroom and after-school activities that DYN offered.

The findings from these two studies expose the hidden significance for organizing civic-minded collective action of activities easily dismissed as recreational or as driven "only" by shared interest.

The Harry Potter Alliance (HPA), one of the case studies underway in the YPP network, offers an example of what pivoting your public can look like on an organizational level (Kligler-Vilenchik, forthcoming; Kligler-Vilenchik and Shresthova 2012). Established in 2005, HPA was inspired by the fictional student activist group Dumbledore's Army, from the Harry Potter

narratives. The group has organized more than 100,000 U.S. fans
to work on political and philanthropic issues such as literacy,
equality, and human rights. Campaigns include "Accio books,
an annual book drive, in which members have donated over
87,000 books to local and international communities; Wizard
Rock the Vote, registering 1100 voters in Wizard Rock concerts
across the nation; Wrock 4 Equality, a phone-banking campaign
to protect marriage equality rights in Maine, and many others"
(Kligler-Vilenchik and Shresthova 2012, p. 11).

HPA has managed to pivot a fan community's energetic iden-
tification with a make-believe world and turn that engagement
toward civic ends. Although members might gather on behalf
of overseas relief efforts or marriage equality campaigns, the
social and creative bond remains a key factor: "I think there's
this balance," reported one HPA chapter organizer. "It's equal
parts making a difference and equal parts meeting more people,
and connecting with people that probably are kinder to them
in a way or just more similar to them" (Kligler-Vilenchik and
Shresthova 2012).

That said, just because someone is your friend doesn't mean
that he or she will necessarily take kindly to the injection of
political themes when he or she is just wanting to hang out. In a
Harvard-based interview study of civically engaged youth, some
young people reported keeping offline civic activities outside
their digital social networks. It is an approach Emily Weinstein
(2013) and her colleagues have called a "fragmented" pattern of
civic identity expression. Others limit their political sentiments
to only a select number of the various platforms they use—a
"bracketed" pattern of civic identity expression. Still others—
the largest of the three groups—integrate their civic views and

participation into their online identities, favoring what Weinstein has called a "blended" approach.

But even the most careful strategizing doesn't always work. Another YPP site is a Chicago-based civic organization that engages high school students in the political process through elections, activism, and policymaking efforts. Interviews there revealed that it's not always easy to strike a balance between socializing and organizing with peers. One young person from the Chicago program confessed that some of her Facebook friends started hiding her in their newsfeeds because they found her political posts annoying; her remarks interrupted the fun social flow (James 2012). Observations like this highlight the extent to which participatory politics entail so much more than acquiring and transmitting information. As has always been the case with grassroots civic organizing, knowing how to read your peer network and edge it forward effectively surfaces as one of the key forms of literacy for politically engaged youth (more on that later).

2. CREATE CONTENT WORLDS

Using inventive and interactive storytelling to achieve public attention and influence.

Last year we saw a sensational example of young people enlisted into a content world to drive cross-generational, transnational political activity: *Kony 2012.*

Kony 2012 is a half-hour, highly produced documentary that has had 97,225,031 viewings on YouTube at the time of this writing. The film was created by Invisible Children, a U.S.-based organization that members of the YPP network have been researching for three years.[1] Founded by three film students from the University of Southern California (Los Angeles) in 2003, Invisible Children aims to "use the power of media to inspire young people to help end the longest running war in Africa."[2] The goal of *Kony 2012* was to make one man famous— Joseph Kony, leader of the Ugandan Lord's Resistance Army—in order to take him down for his crimes against humanity, which included kidnapping children, conscripting them as soldiers, and forcing them into slavery.

It was no surprise when almost immediately after the video started to gain traction, it became a topic of intense debate. In this brief section, I will comment on the ways in which this network of media producers used storytelling to amplify a message and shape an organizing strategy to spectacular and controversial effect. Also noted are less well-known storytelling efforts that add nuance to our understanding of content worlds in the context of participatory politics.

Kony 2012 was not, by any means, an overnight media sensation. It built on Invisible Children's explicit, long-standing strategy of mobilizing public action through storytelling and social media dissemination. Four years before its video sensation, Invisible Children was one of 17 organizations convened by the U.S. State Department to help "combat extremism, . . . [to] better communicate with the rest of the world and to do our job" (Glassman and Cohen 2008). The organization reported 2012 total assets of more than $17 million and has received significant support from entities ranging from right-wing Christian organizations to Hollywood celebrities (Invisible Children Inc. 2012; Kron 2012).

Invisible Children's tactics, said Lana Swartz (2012), a member of YPP researcher Henry Jenkins's team, include "visually arresting films, spectacular event-oriented campaigns, provocative graphic T-shirts and other apparel, music mixes, print media, blogs, and more. To be a member of Invisible Children means to be a viewer, participant, wearer, reader, listener, commenter of and in the various activities, many mediated, that make up the Movement. It is a massive, open-ended, evolving documentary 'story.'" Swartz described these activities as "world-making" in the sense that they contain various points of entry for audiences to cocreate and collaborate in the production of the movement's master narratives.

The logic behind this strategic use of content is based at least in part on the idea of converting attention into action (Zuckerman 2012a). Of the viewers who watch the video, some feel moved to share the link, research the conflict it depicts, or track down Invisible Children and get involved. By creating content worlds that invite newcomers to cast themselves in the story, the movement builds engagement and traction. Although the viral videos are produced by professional crews, other young people can join the effort by, for example, becoming "roadies" who train and then travel with Invisible Children's films to college campuses and other settings across the United States.

"Once you see the story, you want to give something," one roadie told YPP researchers. "And it is easy to pass on the passion of the story and then people want to get involved" (Kligler-Vilenchik and Shresthova 2012, p. 24).

There's another by-product beyond involvement that content worlds can create. Fueled by social media dissemination, stories can take on lives of their own. They can traffic in ideologies that run counter to the best interests of the people with the most at stake in what happens next. While trending at record-breaking rates across leading digital platforms, *Kony 2012* also sparked intense criticism from a range of sources for the world it portrayed. The Invisible Children solution was totally out of step with Kony's current status in the region, argued some African commentators and others, who saw in Invisible Children's call to action a familiar assumption that the solution would originate in the West and deny Ugandans agency in their own fight for justice.

Presented at first as a series of tweets, a critique of *Kony 2012* by the writer Teju Cole (2012) was published in the *Atlantic*. Cole framed the film as a product of what he dubbed the White Savior Industrial Complex, which "is not about justice," he wrote. "It's about having a big emotional experience that validates privilege." That "big emotional experience" is something we need to take seriously as a force within content worlds formed through participatory politics. World-making content is emotional, which is part of what draws people in and makes them enthusiastic, to use Cole's word, to get involved and make a difference.

Nevertheless, as YPP researcher Ethan Zuckerman (2012b) argued in his own widely distributed response to *Kony 2012*, sometimes content resonates and spreads because of the story's simplicity, and simplified stories can, despite good intentions, cause serious harm.

Although I have spent considerable space here on a video that captured global attention, smaller-scale storytelling efforts, supported by tiny budgets and little to no infrastructure, are equally important to consider as we explore content creation as political work—both its merits and its risks.

I'm in Love with Friedrich Hayek is a YouTube video uploaded in 2010 by a young woman named Dorian Electra, with more than 129,000 views two years later. The video opens with a tight shot of John Maynard Keynes's work *The General Theory of Employment Interest and Money*. The camera then pans out to Electra herself—costumed in geek glasses, bobby socks, and saddle shoes—turning away from Keynes and, with delight, toward a stack of books by economist Friedrich Hayek, whose work inspires libertarian thought. Then the music starts. With her confident alto, Electra sings, "Hey, there, Friedrich Hayek, ya lookin' really nice / Your methodology is oh so precise / You break down social science to the fundamentals / Rules and social order are the essentials." The video playfully intersperses cuts of political-economic analysis with shots of Electra swooning over a framed black-and-white photo of the 60-something gray-haired theorist. At one point she plays a game of "he loves me, he loves me not" in Hayek's honor, plucking pink petals off a flower.

In an interview with YPP researcher Liana Gamber Thompson (2012), Electra reflected on her aspirations for the video. By presenting academic ideas in an entertaining way, she wanted to appeal to a range of people, even those who didn't care about politics or agree with her political views. Riffing on the genre of a love-struck, fan-made music video was a way to do that. "I love Katy Perry. I like Nicki Minaj," Electra confessed about her taste in pop musicians. "To have pop music actually infused with something that's a little more substance . . . that gets more people into the dialogue."

This juxtaposition of substance and popular culture is often a hallmark of the content worlds we're seeing in the realm of participatory politics. The right-up-to-the-edge play with identity categories like gender and race can make these worlds both compelling and tricky, especially if our aim is to understand their relationship to voice and influence in public spheres. Assessing the civic and political value of projects such as these requires understanding not only the process that went into making them but also their "digital afterlives" (Soep 2012)—the period after publication when audiences are invited to comment, share, and remix the original medium's messages.

Shit White Girls Say . . . to Black Girls, for example, was one young woman's funny but biting blond-wigged performance of white ignorance. "Not to sound racist," sing-songs the black comedian Franchesca Ramsey at the opening of the video. She then proceeds to act out scene after scene of laughably dumb comments white women have been known to make to black women: "Oh my God, I'm practically black—Twinsies!" "You guys can do so much with your hair!" "The Jews were slaves, too, and you don't hear us complaining about it all the time." That kind of thing. The video spawned lots of knockoffs in which other YouTubers substituted new identity categories—including, for example, the video posted 10 days later, *Shit Girls Say on Their Periods*.

User-generated content streams like these invite more and more media makers into public dialogue, which is a core activity within participatory politics. Yet there's a downside, if your goal is to sustain a critical message: a meme often sets a clock ticking, which before long can turn a clever trope into a tired gimmick, effectively retiring the message whose substance might still deserve serious attention long after the meme has timed out. Achieving "virality" is fun to aspire to but extremely hard to pull off. A problem is that the same mechanisms that enable viral spread—for example, users' ability to riff off of online postings—can also invite deeply disturbing activities such as bullying, savage mockery, and "RIP trolling," in which anonymous commenters harass contributors to memorial pages dedicated to deceased loved ones, a phenomenon that according to researcher Whitney Phillips (2011) has an "accidental politics" of its own.

Creating content worlds involves a whole lot more than making and posting a piece of media. Production of the story, as a meaningful tool for civic and political activity, extends well beyond the moment any one publication goes live.

3. FORAGE FOR INFORMATION

Finding and sharing information through public data archives to discover trends, fact-check, and juxtapose claims with evidence.

In 2008, a group of high school seniors at a Los Angeles public high school used a wireless cell phone–based air-quality monitor to test pollution levels across a range of settings in their community (Niemeyer, Garcia, and Naima 2009). The BlackCloud monitor, as it was called, tracked five environmental parameters: light, noise, sound, carbon dioxide, and volatile organic compounds. Students used the data they collected through the monitors to engage in a series of fictional and real-life activities. They played a video game in which they faced off against an animated adversary using air-quality information acquired through the devices. They produced reports on daily pollution levels. They built "ecotopias" out of wood, nails, and other materials based on their conceptions of model cities for the future.

The big surprise in their findings was that among all the sites where they tested air quality, including those notorious for environmental contaminants (e.g., the local dry cleaner), among the worst was the very place that brought them together to do this work: their classroom. Reflecting on the outcome of this activity, BlackCloud's developers said that the experience enabled students to generate and disseminate new information about their local neighborhoods. The young people established "knowledge-based agency within their communities using digital media both

for real-world data-acquisition and real-world communication"
(Niemeyer et al. 2009, p. 1076). BlackCloud partner and study
coauthor Antero Garcia, in whose classroom the investigations
took place, saw within this work the beginnings of social impact:
"We can start thinking about aggregated change as a result of
incremental change."

I'd like to frame this project as an instance of information
foraging, the third tactic of participatory politics. Young people and their mentors utilized emerging tools and platforms to
discover, organize, and share untapped data, with the goal of
instilling a sense of public responsibility and identifying solutions to environmental harm. Information foraging upsets the
conventional categories of who normally collects, fact-checks,
analyzes, and reports on original data. Foraging implies that not
only accredited experts are in a position to glean public assets;
resources reside all around us, and there shouldn't be a million
obstacles, nor should it require a million dollars, to access those
resources. When you think of foraging, you might most readily
picture someone picking blackberries from a bush in a park or
herbs from a plant alongside the freeway. The same concept can
be applied to gathering data that are hiding in plain sight.

Two developments at the intersection of technology and
culture heighten the role of information foraging as a tactic of
participatory politics. The first development is the emergence
of the social networks that young people increasingly rely on as
repositories of knowledge. The BlackCloud work was carried out
in 2008, relatively early in Twitter's adoption, especially among
teens. Even at that early stage, students who were part of the
experiment used Twitter to broadcast information and encourage environmental stewardship among their friends and followers. By now, we might have a hard time conceiving of civic
activities that don't in some way engage digital social networks

(Middaugh 2012). But it's important to mark the salience of these platforms specifically as a means by which to circulate information between elites and peers on matters of civic significance. YPP research reveals the striking extent to which young people depend on participatory channels for information. Howard Gardner and Carrie James's research group (Gardner, James, Knight, and Rundle 2012) interviewed young people who knew that their friends relied on them to follow current events; one reported that peers turned to her Twitter feed daily to follow what was going on in the world, and she took that role seriously, almost like an assignment (James 2012). In Cohen et al.'s (2012) national survey, 45 percent of the respondents reported getting news at least once a week from family and peers through social media feeds—nearly as many as those who consulted newspapers or magazines.

It's energizing to imagine all the information young people can now readily access through online searches and custom channels they engineer every time they add or drop a social media connection. But it's also daunting, and perhaps a little scary, to contemplate what it takes to make sense of that waterfall of data. Young people, it seems, share these concerns. When asked if they and their friends could benefit from learning more about how to tell whether news and information they find online is trustworthy, 84 percent said yes.

The second development relevant to the role of information foraging in participatory politics is the phenomenon of "big data." By now you may very well be tired of hearing this term, if you are among the developers, researchers, and marketers who in the last few years have become deeply interested in its potential. *Big data* refers to new forms of digital information gathering meant to reveal unexpected patterns that can help us understand, navigate, recalibrate, and/or monetize our complex social

world, for better or worse (boyd 2010). Although the term is at
risk of overpopularization and its definition can be hard to pin
down, the spread of data collection as part of nearly everything
we do—from loving to banking to spectating to learning—has
undeniable effects on how young people do politics as well.

BlackCloud is probably best understood as a project based on
"small data"—that is, information collected on a finite sample in
manageable volume for a specific, targeted purpose. And yet it
hinted at what was around the corner, in 2008, when reports on
the project were published: a burgeoning interest in opportunities for citizens, including nonexperts, to feed, assess, and share
data on a mass scale (and in the process develop expertise).

One of the key developers of the BlackCloud project, Greg Niemeyer, moved on to join the Data and Democracy Initiative, a
University of California–Berkeley project with academic, government, and industry partners intended to facilitate deliberation,
understanding, and rapid mobilization through data streams
flowing from digital and social media tools. Competitions sponsored by backers, ranging from the Mozilla Foundation to the
National Science Foundation to the Federal Communications
Commission to the White House itself, have invited citizens,
including youth, to propose technology projects that make, in
the words of the Apps for Communities call for entries, "local
public information more personalized, usable, and actionable
for all Americans."[3]

Big data itself has become so big that there is even a kind of
metaproject designed to make the concept more publicly accessible. The Human Face of Big Data, a group that has a special
program called the Student Face of Big Data for children and
teens, is "a globally crowdsourced media project focusing on
humanity's new ability to collect, analyze, triangulate and visu-
alize vast amounts of data in real time."[4]

BlackCloud took place in the context of classroom-based environmental education. The project was a case of what is known in that field as citizen-science, referring to efforts in which nonscientists—in this case, youth and their allies—frame research questions and collect and analyze data. Journalism is a second field in which information foraging through digital and social media has had a serious civic effect. The citizen-hyphenated version of journalism refers to projects in which individuals not formally trained as reporters investigate, document, and broadcast the news.

It is beyond my scope to account for youth involvement in citizen-journalism. In fact, I'd like to distinguish that sphere of activity from the field of organized youth media programs, which have made special use of information foraging as a tactic to investigate and raise awareness of issues relevant to freedom and equality. Youth journalism projects do not seek an audience for untrained young reporters. Rather, they educate young people in the core tenets of journalism while partnering with them to bring next-generation techniques to the work of reporting about and in public spheres.

I work at one such project, Youth Radio in downtown Oakland, California, which is where I first met Pendarvis Harshaw, whose story opened this report. At Youth Radio, young people—the majority are youth of color from low-income communities—collaborate with professional producers and editors on content for outlets including National Public Radio, American Public Media, the *Huffington Post*, *National Geographic*, *GOOD* magazine, iTunes, and local and commercial public radio stations around the country (Soep and Chávez 2010).

Some of the organization's most ambitious work in the last few years has come from its investigative desk, including two series that were recognized with major national honors (one with the Robert F. Kennedy Award and one with the George Foster Peabody Award). In *Navy Abuse*, Youth Radio investigated a U.S. Navy base in the Persian Gulf where a culture of hazing and abuse targeted a young gay sailor and others in the unit. In *Trafficked*, the newsroom dedicated several months to tracking the social and policy dynamics that drive underage girls into commercial sexual exploitation and the legal system that criminalizes teens who have been trafficked and makes it exceedingly difficult to prosecute their abusers.

I was a member of the production teams for both of these projects and have previously written about them (Soep 2011, 2012). I am struck by the ways in which digital and social media revealed information that was pivotal to each story's production and dissemination. In *Navy Abuse*, at one point the fate of the story hinged on whether the reporter could verify the unit's chief petty officer's whereabouts and the fact that he had been promoted after abuse allegations came to light.

To tell a fair story, Youth Radio needed to make every effort to give this man an opportunity to share his account of what had happened on the base. But we couldn't locate him—until a source told the youth reporter about a social media site for military personnel and veterans. That Web site provided crucial information about the chief petty officer's rank and deployment status, which finally enabled Youth Radio to get official confirmation of the facts and to send questions directly to him.

Digital and social media channels also drove dissemination of the story. Online, the newsroom published original documents procured through Freedom of Information Act requests and other materials that couldn't be shared on a radio broadcast. The story's main character experienced the effects on his life of going public with his story on many levels, including his

Facebook identity. After it aired, he realized he could no longer use Facebook as a place to socialize and relax. His new persona as a spokesperson on the effects of the U.S. military policy of "Don't ask, don't tell" meant letting go of his more carefree online identity, and he had to turn his Facebook profile into yet another tool for that work. It was not without mixed feelings, it seems, that he exchanged his profile photo for one that seemed more appropriate to the context.

In the second Youth Radio investigation, *Trafficked*, digital and social media emerged as central forces in the story itself. The reporting team uncovered a network of local photography studios and public relations consultancies that had sprung up to help clients produce digital profiles for Web sites that sold girls for sex. Moreover, efforts to crack down on prostitution sites— a move some researchers say might erase traces of perpetrators' activities that are actually useful to law enforcement (boyd, Thakor, Casteel, and Johnson 2011)—were an important backdrop to Youth Radio's reporting and showed up through heated comment streams in response to the story after publication.

In both of these examples, young reporters and their colleagues deployed digital and social media tools to forage for crucial information while reporting on how those very tools were factors in the stories. The producers then exploited the same tools to spread the news. In science and in journalism, we are seeing more and more examples of young people unlocking information to advance public awareness and facilitate concerted action. That said, as usual, it's not all good news.

No one wants unflattering information to surface unexpectedly, and the dynamics can get especially intense when multiple power discrepancies are in play. In one YPP study of a large urban school district's youth advisory committee, researcher

Margaret Rundle, part of a Harvard team (Gardner et al. 2012), learned about students' efforts to have a role in teacher evaluation. In a meeting the youth group had captured on video, the superintendent had, in their view, expressed support for these efforts. But when the group went public with that endorsement, the superintendent backed off and sought to distance herself from that position.

The fact that the young people had video documentation of the original meeting certainly strengthened their position, but it also made them a greater threat. Information is power, but it can also get you in trouble. Adult mentors, colleagues, and allies need to be prepared to support young people as they seek and expose information that elites prefer to control, especially when young people are accessing sensitive records and could later be accused of doing something wrong to get them. We need to make sure that they know how to do it right, within the bounds of the law (even if they choose to violate the law), and that they're prepared to stand up to possible retaliation.

Finally, there is something potentially misleading in the concept of foraging that I'm offering here. Information is newly available as a result of big public digital databases and other social and searchable archives, but when that information reflects unfavorably on powerful people, utilizing it can require sophisticated knowledge of everything from computer programming to statistics to design principles to techniques for sweet-talking or bamboozling one's way into networks that have every reason to block youth.

The veneer of transparency can be more dangerous than obvious efforts to distort or block access to information, because it makes institutions that are actually keeping secrets appear forthcoming. "Beautiful data can be seductive," after all, by "giving the illusion that if we only have enough information we will be able to make the right decisions" and by misleading us into thinking that data contains answers rather than new questions (Grant 2012).

As I more fully explore below, what Wendy Hui Kyong Chun (forthcoming) calls the "politics of storage" assigns value to networked data to the extent that online activities leave a trace and connect to others. Information foraging can be used by youth and also against them, leaving traces they might not recognize until it's too late. Moreover, for every government- or corporation-backed invitation for citizens to come up with tools that mine public data stores, there are stepped-up efforts to keep proprietary data off-limits and under administrators' control. Young people's engagement with participatory politics entails tracking and utilizing developments on both sides of this dynamic.

4. CODE UP

Designing tools, platforms, and spaces that advance the public good.

In 2010, Youth Radio launched a new arm of its production company, the Mobile Action Lab.[5] The lab partnered young people with professional designers and developers to create mobile apps. Motivating the work was the realization that it was no longer enough for young people to create content using existing tools or to deliver their stories on the available platforms. They needed to be the ones who were engineering those tools and designing those platforms, which increasingly determine who knows what, how information circulates, and what sparks change.

Sounds good, right? We thought so, too. But launching the lab turned out to be more challenging than any of us realized.[6]

Code switching wasn't just a metaphor anymore for the kind of work that happened every day at Youth Radio. Young people had always been called on to shift how they expressed themselves, based on which media audience they were addressing. Now they had to learn to switch from content creation to software development. And—no small matter—they had to learn to code.

Members of the Mobile Action Lab applied what they knew about writing and editing radio scripts to the process of learning computer programming languages and updating civic software. They got used to the idea of engaging users rather than addressing audiences. They learned to think in terms of a "minimum viable product" (MVP): the simplest, most stripped-down version of what you're trying to build. They forced themselves to release an MVP to users as quickly as possible, inviting early adopters to help make the product better. Young people figured ways to make their designs intuitive and their user experiences social, pitched their projects to anyone who would listen, created systems to track testing and optimization efforts, and debated how much good any given app had to do for it to qualify as "serving the community." (For example, was it enough to say that the app is fun and therefore a relaxation mechanism for stressed-out youth? I didn't buy that one.)

At first we set out to create technology that addressed community needs. Eventually we modified our approach, abandoning the flawed assumption that our job was to churn out tools to fix local problems or make up for shortcomings. Now the Mobile Action Lab seeks to create apps that spark storytelling and citizenship, often by stoking community momentum that is already building around an issue or opportunity.

Youth Radio's Mobile Action Lab is part of a growing movement to engage young people in efforts to design software that supports transparency, democracy, civic engagement, and justice. Youth App Lab was another early effort along these lines, founded at Youth Uplift in Washington, D.C., by engineer Leshell Hatley, who wanted to create pathways for black youth into computer science. Iridescent is a national program that builds technology literacy in girls, in part through the Technovation Challenge, in which the girls learn to produce mobile apps and launch start-up companies. By exposing students to computer science and technology, the San Francisco–based Black Girls Code set out to increase the number of women of color in the digital space and enable them to be leaders and builders of their own futures. The Hidden Genius Project in Oakland strives to unlock pathways for young black males into careers in software development and design.

Some of these efforts and others drew early inspiration from Apps4Good, a British-based project that was perhaps the first to engage teens in app making. Many used tools like App Inventor, built by Google Labs and now run out of MIT, which enables people with no computer science training to create apps. And many of these organizations are now in the process of updating not only the apps in their portfolios but also the structures and scopes of their own programs. They're graduating from start-up mode into periods of establishing consistent and sustainable frameworks for engaging youth and communities through mobile software design.

The strategy of coding up can be seen as part of an even larger movement taking hold in education: the Maker movement, of which Dale Dougherty is credited as a founding force. In 2006, Dougherty held the first Maker Faire, a grand festival billed as "the greatest show-and-tell on earth" that drew more than 150,000 enthusiasts in 2013 and has spawned more than 50 community-based Mini-Maker Faires around the world. Though best known for these massive family-friendly events that showcase technology-infused do-it-yourself (DIY) culture mixed with craft, innovation, and playfulness, the leaders within the movement are increasingly working to create "Maker spaces" in public schools. In 2013 the White House convened key figures from the Maker movement, hoping they held a key to revitalizing science, technology, engineering, and math education and even the future of the U.S. manufacturing industry.

The relationship among the Maker movement, learning, and civics is not without tension, however. We still have a lot to learn about what triggers some young people to get started as makers, what marks trajectories of advancement, and what framework has the potential to connect informal and formal learning domains (Sefton-Green 2013). Dougherty recently came under some criticism for accepting funding from the U.S. Defense Advance Research Projects Agency (see Dougherty 2012 for his response to critics). Makers themselves, though deeply techie in sensibility, have voiced the concern that software development alone, without the integration of tactile materials from the physical world, doesn't fully utilize the learning potential of DIY Maker culture and therefore shouldn't necessarily be included as fully representative of the movement's core values. "We have," says Julian Sefton-Green in a recent literature review on making and education, "inherited an idea of creativity as imaginative and organic, whereas we still tend to think the digital pertains to the age of the machine" (2013, p. 56).

Finally, calling it a movement at all strikes some as problematic, to the extent that communities—in particular working-class ones—have forever been "making" useful and intriguing objects in a DIY spirit and are now sometimes targeted for outreach from

Maker movement institutions without real acknowledgment of their leadership.

All that being said, implicit in coding-up activities are the beliefs that young people are makers and not just users of technology and media and that their activities and products have the potential to bring good to their own lives and communities. We are in the early years of seeing this work build momentum, and the challenges are significant. As any maker knows, creating a prototype can happen fast, but building a friction- and glitch-free project that totally delivers on its promise and gains runaway traction with users is rare, and takes time, luck, stamina, and money. Along the way is plenty of failure, which start-up types like to celebrate but which can be deeply demoralizing for young people whose humanity and intelligence are already under assault and for organizations whose existence depends on cheerful grant reports touting success.

You may note another theme in "code up" initiatives: a consistent interest in building technology for good and creating pathways for young people to higher education and viable careers. This is at best a potent combination that sees citizenship as both a concern for justice and the pursuit of a life that includes meaningful and sustainable learning and work. At worst, young people never get to reflect critically on the potential tensions between these two orientations, and youth can be included in token exposures that can be hard to utilize for any real effect on their own development or on life in their communities.

There's a lot to learn from youth app development and maker spaces about new models for mentorship or collegial pedagogy between novices and experts (Grossman, Chan, Schwartz, and Rhodes 2012; Soep and Chávez 2010), because there is a huge gap between what it takes to create the simplest app, robot, e-textile garment, or digitized contraption, on the one hand, and the work required to bring a sophisticated product to market, on the other.

Finally, standard metrics like download counts are useful in the realm of civic apps, but they fall far short of capturing nuanced measurements of the effect of these projects. We are only beginning to frame analytics that assess the quality of product design, learning experience, and outcomes for the makers and their communities while enabling the developers to maintain independence and agency with their codes, contents, and crafts.

Projects like the Mobile Action Lab, Black Girls Code, the Hidden Genius Project, and maker spaces that are cropping up around the country are designed to engage young people for sustained periods, like months or even years, in technology projects that do good in the world. There's a second phenomenon that fits with the tactic of coding up and that conforms to a very different pace: a caffeine-charged, sugar-rushed, amped-up, around-the-clock-and-then-we're-done mode of civic engagement. I'm talking about the community hackathon.

You arrive with your laptop by 8:30 A.M. on a Saturday to a nondescript conference center or office space emptied of its weekday employees. Bagels, doughnuts, sticky Danishes with glossy neon centers, and cardboard boxes of coffee cups fitted with plastic nozzles are laid out on tables lining the walls. Maybe you sign in. Maybe you get a T-shirt printed with a slogan like CODE FOR OAKLAND or HACK IS NOT A FOUR-LETTER WORD. You gather in the auditorium and say who you are: a coder, a designer, or maybe a local person with an idea for a project. You clap for the keynote speaker. You gather with your assigned group in a designated corner of the space. There are lots of sticky notes and butcher-paper pads and makeshift circles of people sitting cross-legged on the carpet. People talk fast, brainstorming ideas for what you'll build together over the next 12 hours or so. Someone is scanning app stores and Web sites to see what's already out there in the market, and someone else is writing down what others are saying, drawing possible graphics and home screens, and listing possible names for your project. One or two people are already coding.

Before you know it, it's getting dark outside and the breakfast goods have been pushed aside, making room for greasy-salty options like pizza and potato chips, and soon you're slamming together a presentation deck with slides that tell your story and make your case about how this piece of technology will transform this community. You're deciding who will say which parts and how to force-fit the spiel into the strict five-minute time slot, and all this time some of you are still frantically coding so you can demonstrate an actual clickable prototype of what you purport to build. You still can't agree on the name, but time is running out, so you pick one and sip your tepid coffee and wait to be called onto the stage. Judges judge, winners are announced, prizes are handed over, and plans are made for the next steps on the projects conceived at this event and for hanging out late into the night with the friends you've made.

I attended my first community hackathon when Youth Radio debuted one of our apps—a food-sharing platform—at one such event in Oakland sponsored by Code for America. These events have been happening for some time, originating out of developer communities and even making a cameo appearance in *The Social Network*, the movie about the founding of Facebook. Andrew Schrock, who works with YPP researcher Henry Jenkins, is studying hacker spaces and traces their roots to mid-1990s, off-the-grid gatherings in places like shopping mall food courts, where young coders would often bring a punk sensibility, a cliquishness, and an ethic that worshipped technical expertise. The law, Schrock says, was not always so important. What started as unauthorized gatherings of people "identified only by nicknames like 'Deth Vegetable'" have emerged, over the last couple of years, as official, underwritten events aimed at bringing together techies and communities to cocreate applications that improve lives (Schrock 2011).

Where do young people and the social good fit in this mix? It's a question many hackathon hosts are actively trying to answer, not always gracefully. There are two main ways that young people tend to show up at these affairs. Sometimes they are invited to present the community needs that techies are being enlisted to fix. This points to one of the tensions within the hackathon model. It can reify tired dynamics that frame youth as authorities only on problems that mire their lives and developers as saviors who can swoop in with solutions (and then return to their day jobs once the prototype is created).

Actually, hackathon organizers who go out of their way to engage young people with direct knowledge of what's most needed in their local communities are a lot better positioned for success than those who simply speculate about what they might do to improve something they have not experienced and do not understand. Code for Oakland, for example, spent time before its hackathon meeting with local groups and learning about the challenges residents were facing that could be addressed through technology. Moreover, developers who take time out to lend labor and talent to neighborhood coding efforts deserve recognition for stepping up. That said, by now even hackathon enthusiasts are seeking new models that engage various sources of knowledge relevant to community affairs. They are exploring ways to launch sustainable projects that advance beyond one-off designs, which rarely evolve into fully developed tools with a measurable effect on local people's lives.

There is, of course, a second way that young people participate in hackathons: as hackers. Although I haven't found data on the average age of hackathon coders, many, if not most, would still count as "youth" as defined by a typical community-based organization. What strikes me here is the extent to which hackathons can reflect the inequalities evident in technology writ large. Some young people are there to testify about the best interests of their communities, and some are there to build software solutions. The cultural capital and earning potential associated with these two modes of youth participation are by no means equivalent. In this sense, the interventions described earlier, aimed at teaching young people how to design and code, emerge as key to achieving greater equality in participatory politics. Voicing one's concerns and expressing influence in public spheres increasingly requires not only the use but also the development of digital tools. Knowing how to create those tools, products, and environments isn't simply a set of technical skills. It's a mandate for civic learning.

Whether because of funding structures that require it, philosophies that value it, or a combination of the two, coding up as a tactic of participatory politics often favors open-source development, which involves surrendering some degree of control over the products being created. If you release the code through repositories where others can find it, change it, and improve it, you'll get better technology and enrich public spheres.

In my own experiences working as a producer with youth making media and technology, I've found this point of view highly persuasive. But I want to close this section by sharing a question I don't know how to answer. In 2012, I was working with a group of young people at Youth Radio on an app called Forage City, a food-sharing platform that invites users to distribute excess fruit bounties to neighbors and nonprofit organizations. "Uniting citizens of leftover nation" is the project's tagline.

This was shortly after we had released a beta version of the app at the Oakland hackathon. Based on community feedback and user testing, we came up with a set of design modifications we wanted to make to improve the user experience and expand the app's appeal. We also got excited about the idea of framing Forage City as an open platform rather than a locked-down application. We had already planned to release the source code, but now we were considering an approach that invested less in the idea of a full-service app and more in a flexible set of components that users (including fellow developers) would be invited to remix as they saw fit for their specific communities. Sharing economies aren't limited to fruit, after all, or even food, for that matter. And our design was just one design. If we open-sourced it and explicitly promoted it to third-party developers, we might see multiple custom iterations pop up all around the country, with labor, credit, and positive effects shared by a range of cocreators.

I was busy extolling the virtues of this approach when my 17-year-old colleague interrupted. If we open up the platform like that, he asked, how do we make sure the people we care most about are the ones who get to use the app? It's people in the hills, he said—the wealthier parts of Oakland and the East Bay—who will have the resources to get into the code and turn it into what they want. The people in the flats—the more economically depressed areas—won't. So all that healthy food will stay in the hands of people who can already afford to buy it at Whole Foods. And people who are struggling won't be any closer to the goods they need.

In this interaction, I would argue that my colleague was "doing" participatory politics on a micro level, by challenging me for missing the hidden pitfalls of embracing a give-it-away approach to code without fully grasping our continued responsibility. I'm not sure I gave him a satisfactory answer in that moment (or that I have one even now), but I do know that I came away with renewed appreciation for the moment-to-moment work of having a voice and expressing influence in public spheres. Sometimes it doesn't happen in front of a big audience at an orchestrated event or a product release. It can entail challenging your own colleagues and pushing them to keep the public good at the center of the activity you're carrying out together. Even when the activity is developed through computer code, the politics are also flowing through face-to-face conversations that outside parties might never hear but that have everything to do with the design and effect of what's released into the world.

5. HIDE AND SEEK

Covering tracks and protecting information from discovery as actors engage in politics that only selectively emerge into public awareness.

On November 11, 2011, two Latino students in their early 20s walked into a border patrol office in Mobile, Alabama, hoping that they wouldn't freely walk back out. "Last words?" one asked the other, as they fired up the cell phone camera in the car and prepared to step inside. The video of their encounter with two white border patrol officers has the now-familiar look of an engineered confrontation "caught" on amateur video (it's unclear from the footage whether the people on camera knew they were being taped).

From the point of view of one of the young men, you see a nondescript office space, with the requisite flags, framed head shots, and couches lit from above by overexposed rectangles of fluorescent light. He tells a woman who greets them that he and his companion are lost. "Hang on a second," she says, sounding nervous. The woman slides her security card through a reader, unlatching a door, and exits to find help.

Two officers, one in uniform, emerge from the other room. "Hey what's going on? How you doing? Can we help you with something?" one asks.

"Yeah, you know what?" one of the students responds. "I'm actually not lost. I'm just kinda pissed off. What are you all doing here?"

"Doing our job. Why?"

"What's your job?"

"To enforce immigration laws."

"That what you do?"

"Yeah, that's what we do. What's it to you?"

The young man with the camera says, "I'm illegal, too."

"Oh you're illegal?"

"So you think I should get deported?"

The officers ask for an ID. The one in uniform looks at the card, flips it over to examine the back, and asks, "How'd you get to the United States?"

"'Cross the border."

"When did you do that?"

"Long time ago."

The following words flash on the screen: "After the cell phone signal dropped, Jonathan and Isaac were detained and transferred to the Basile Detention Center in southern Louisiana. Inside the detention center, they're meeting many people like them—immigrants who've committed no crimes. The administration is lying when they tell us they are only deporting serious criminals."

YPP ethnographers Arely Zimmerman (2012) and Sangita Shresthova are considering this video as part of their research on undocumented youth and their allies' use of participatory media to oppose U.S. immigration policy and advocate for the DREAM Act, legislation that grants conditional legal status to college students who were brought to the United States before they turned 16. This tense scene of entering a space of policed authority is a trope as prominent within activist videos as the arrival stories anthropologists use to introduce their ethnographies of faraway cultures (Pratt 1991).

In this particular video, Jonathan and Isaac come across as somewhat scared, as though maybe they hadn't fully thought through what they were about to do when they hatched the plan to turn themselves in to border control. But like many documentary-style confrontations, this video has a mix of genuine and dramatized affect. In a subsequent interview, Jonathan told the Web site CultureStrike, "We went undercover and decided to *pretend* we were afraid, *pretend* we are not connected in any way" (emphasis added), as part of a strategy that would get them sent to a detention center, where the two could continue their organizing from inside (Chen 2011).

A follow-up story on the Web site ColorLines reported, "The lesson, they believe, is that undocumented immigrants are safer when they come forward and organize instead of cowering in the shadows. It's there [inside the detention centers] that ICE [Immigration and Customs Enforcement] does most of its enforcement work, they say. It's there where it's impossible to hold them accountable" (Hing 2011).

At first glance, this video would seem to be consistent with the other cases of participatory politics I've considered above. All four sets of tactics—pivoting public discussion toward political ends, creating content worlds designed to instigate action, foraging for information, and developing digital tools to voice one's concerns and exert influence in public spheres—center on proclaiming civic positions.

But what's striking in Jonathan and Isaac's story is the interplay of disclosure and cover, voice and silence—activities that take place in the full light of public awareness and those that happen in the shadows. This is what the writer Alexis Madrigal (2012) calls the "dark social" corners of digital life—spaces that are hard to find and track. The power of this video, and the larger phenomenon of youth coming out as undocumented, without papers and at risk of detention and deportation, resides in the tension these public gestures create against the backdrop of a larger expectation of secrecy and silence.

In studies of politics in the digital age, we tend to focus on speech—new ways that technology enables overt, amplified, and ever escalating civic expression and action. We would be remiss not to account also for the tactics young people deploy to mix authorship and anonymity, vocalization and silence, especially under digitally enabled conditions of heightened surveillance.

"The re-conceptualization of the public sphere around silence, instead of speech, provides the tools necessary for grasping the political significance of anonymous speech," said political philosopher and YPP researcher Danielle Allen (2010, p. 108). The conceit of anonymity, if not its reality, is of course a hallmark feature of digital identity. Public spheres are certainly made up of rituals and mechanisms that foster discovery and disclosure, but there are also proliferating ways to close off, cover up, and disguise certain kinds of conversation. Silence doesn't necessarily mean the status quo has won. Tactical silence, Allen argued, can have important political value as a destabilizing and deceptive force:

[I]f powerholders take silences as affirmative or acquiescent when in fact they are negative and resistant, powerholders will develop significant misperceptions of the realities they inhabit until the misalignment between their perceptions and reality becomes so great as to reach a breaking point, and their capacity to act in the world, their power, simply gives way. Silence, or fake acquiescence, can serve as a political weapon when it is used to mislead powerholders about the truth of their situation; not knowing the truth of their situation, they will fail to make sound practical judgments about it. (p. 9)

YPP researcher Ethan Zuckerman (2008; forthcoming) and his colleagues have examined striking cases of digital activists around the world expressing resistance by masterfully (and sometimes hilariously) outsmarting censorship tactics like keyword filters that governments use to block online posts that contain terms like *human rights* and *democracy*. Within the United States, organizers who are targets of intensified state-backed suspicion—including, according to YPP research, undocumented youth activists and Muslim-American youth organizers—have found their own ways to play hide-and-seek inside systems of surveillance. It's a game without clear-cut rules.

YPP research director Sangita Shresthova interviewed participants in an incident at the University of California at Irvine in 2010, when a group of students associated with the Muslim Student Union disrupted the speech of Israel's ambassador to the United States during his visit to campus (the students were later found guilty and their club suspended). One interviewee confessed that she found it difficult to figure out what to reveal about herself and when to keep silent.

"It's hard," she said. "I try not to post too much personal information just because you don't know who—I'm sure there are people that don't agree with my viewpoint that are friends with me on Facebook. . . So you don't want to post too much personal information . . . I mean, we [get] death threats and stuff, hate emails and stuff. . . . Like one of the things I try not to do is to post like where I am . . . like physical location. I try to limit things."

Limiting things is part of what Sunaina Maira (2011) described as a larger project of Muslim-American youth in her study on what politics are "possible or permissible" under the "War on Terror" (pp. 2–3). She distinguished between surveillance *effects*, or strategies that chill dissent, and surveillance *affects*, which "mediate the production of selfhood in a period of permanent surveillance, where the self is constantly performed in public view" (p. 13).

As with the YPP case study of undocumented youth, here we see an intriguing and sometimes counterintuitive dynamic between hiding from view and stepping out. Maira discovered that one way Muslim-American youth dealt with the perceived stigma of government profiling or surveillance was to reframe those activities as achievements. She quoted one young man who told her that "it wasn't like a badge of shame, it was like 'Yeah, the FBI is listening into *my* house'" (p. 16).

Likewise, Shresthova has seen Muslim-American youth and DREAM activists alike embrace hypervisibility. Although organizers sometimes revert to face-to-face interactions to plan events or discuss charged topics because of the likelihood of tracking social media posts, others opt to put it all out there, as they say. "There's [*sic*] no victims here, and we're not going to crawl into a hole" is how Shresthova characterizes the way some youth groups respond to the pressure to self-censor, based on her team's research. For the DREAMers, she said, there's a belief that "if you're under cover, if you're hidden, you're less powerful than if you're visible. When deportation orders come, if you're visible and out there, there can be a rallying around you. If you're not known, you're vulnerable."

Managing visibility and invisibility, speech and silence, is itself a participatory activity in the digital age. Under the old system of institutional politics, established gatekeepers were better positioned to control the flow of information, delineate the conditions in which dialogue occurs, and determine which people's identities and activities are revealed and concealed. In an era of participatory politics, elites continue to play important roles in all of these areas. But, as YPP researchers have argued, institutions operate alongside young people and their peers, who actively pursue, analyze, and critique information about issues of public concern; shape the creation and flow of news; mobilize others through social networks and organized groups to accomplish political goals; and help decide what information enters the public record and what stays unattributed and hidden from public view (Kahne et al., forthcoming).

Literacies That Support Participatory Politics

For the five tactics I have presented here, the trick, of course, is knowing how to utilize these activities in ways that achieve the desired effects on issues of public concern. It's one thing to name some tactics young people are using to have a voice and exert influence on public affairs. It's another thing entirely to create compelling ways to organize communities around these kinds of activities, meaningfully and equitably.

That brings me to literacies. What are the forms of know-how that power participatory politics? It is beyond the scope of this report (and of my own know-how) to offer fully developed curriculum models for promoting participatory politics. My hope is that the discussion here will become part of a larger set of efforts great educators are working on to imagine and pilot new ways to advance digital civics in its various forms among youth. The YPP network itself is committed to partnering with educators to pursue this agenda in its next phase (anticipated to be 2013–2016). For our purposes here, then, my aim is to identify some of the emerging literacies that seem most relevant to the tactics of participatory politics, with some exploratory ideas for how they might be cultivated through learning activities.

Literacies are best conceived as practices honed through participation and situated within social contexts rather than as discrete, transferable skill sets. Think verbs, not nouns, and imagine collective orchestration versus individual knowledge acquisition (Heath, Flood, and Lapp 2008; Street 2001; Varenne and McDermott 1999). The rising salience of participatory politics forces us to rethink both core literacies and our conventional ways of teaching (DeVoss, Eidman-Aadahl, and Hicks 2010) and offers a useful starting point for educators seeking to build learning environments that spark civic engagement.

Literacies are key, because while participatory politics tend to be equitably distributed across different racial and ethnic groups, and youth engage in participatory politics about as often as they do institutional politics, the majority of youth are *not* engaged in participatory politics (Cohen et. al. 2012). In her interviews with high school students around the United States, YPP researcher Chris Evans (personal communication, 2012) found that most don't automatically come up with ways to utilize digital and social media to advance the social good, although they are able to do so when prompted, conveying what Evans called a *digital imagination* that outstrips their actual engagements.

Moreover, young people who aren't deeply involved in civic projects or especially media savvy seem more likely to use digital tools in relatively lightweight ways, such as circulating information by forwarding a link, rather than through more ambitious and challenging activities like producing original content that raises awareness and mobilizes others to act (Cohen et al. 2012).

Jennifer Earl is working on a study as part of YPP that examines the efficacy of radically low-labor gestures like forwarding a link or "liking" a slogan. These are activities she called *flash activism*. We've seen their potential for effectiveness in events surrounding the groundswell of protest against antipiracy legislation in the United States, which young people followed closely (more closely, in fact, than presidential election news, according to Pew research), mobilizing online friends and followers to oppose bills in Congress that they believed would limit free speech and innovation.

And so by no means should we dismiss the value of flash-activist forms of civic engagement without knowing a lot more about their role in social change. But if equity means that citizens "take turns" achieving political gains and accepting political losses while also honoring the gains and losses of others, as Allen (2012) has argued, then we are well served as a society to build learning environments that prepare young people for the full range of opportunities at their disposal to engage with and remake public spheres.

Let's start with the literacy demands behind our first tactic, pivoting your public, which entails mobilizing (apparently) latent civic capacity within networks that originate in popular culture. In order to activate their peers toward political ends, young people need to know how to feed their social networks and forecast the ways in which their activities in the present will play out in any given project's digital afterlife. To maximize this tactic, young people need to understand the tacit etiquette, and build the stamina and habits, that undergird the networks poised for mobilization. Standards of reciprocity are rarely spelled out within the terms and conditions of Web sites or mobile apps. And yet mastering those protocols can make all the difference for young people who reach the point of wanting to use the power of their networks to raise awareness about a social or political issue.

Pendarvis Harshaw's highly active and receptive community of friends and followers provides a model of an informal network that is well fed, well held, and poised to share. Again and again, however, within YPP research and in our own networked lives, we have seen clumsy pivots. Someone misjudges the social dynamic and introduces an issue in ways that inspire eye rolling or outright resentment rather than productive action.

Educators seeking to help young people get smarter about pivoting their publics might start by collecting cases of efforts that were wildly successful and some that were flaming disasters, and to identify what features and design principles distinguished the two. There are thorny ethical questions that could spark meaningful discussion as young people work to develop literacy in this area. Is it okay to fake interest in other people's work or to contrive token gestures of digital solidarity if the real intent of these moves is to set yourself up for reciprocal support for the efforts you care about?

There is also the crucial matter of dissent. How can young people galvanize a community with shared popular-culture passions, for example, without alienating those whose political views don't line up with the majority or the most vocal within the group? Of significance here is not just how you pivot your network toward issues relevant to public affairs but also how you effectively pivot back.

The second tactic, creating content worlds, involves the use of inventive and interactive storytelling to achieve public attention and influence. This can call for some very specific technical skills. To create compelling media that trigger concerted action, young people need to know how to plot, cast, and enact story lines that translate issues and arguments into provocative narratives that enlist others as coproducers. There is tremendous value in knowing how to record and edit media in sophisticated ways.

But content worlds don't always require advanced technical skills. As important are platforms that make it easy for a community to put up a quick Web site or polish snapshots captured by cell phone, as well as learning environments that support the rhetorical skills of conversational storytelling, cultivate the drawing skills to create posters for rallies, or craft skills to make masks and costumes. Above all, content worlds thrive on curiosity and the conceptual capacity to superimpose popular culture onto the political realm, often through remix and appropriation.

There's nothing particularly new about this core set of activities. What's remarkable is the increasingly important role the activities play not as expendable extracurricular talents but as capacities that are essential to active citizenship. It's much easier today than it was ten or even five years ago to acquire the necessary equipment and technical knowledge to carry out these tasks. Today, 31 percent of teens ages 14–17 have smartphones, according to Pew Internet and American Life data (Lenhart, 2012). That's how many carry around production and distribution platforms in their pockets at all times, to say nothing of those who might access those technologies through public libraries or schools or by borrowing devices from older friends and family members (whose rates of cell phone ownership are even higher). And production values are much more forgiving today than they once were, with the growing appeal of gritty, home video–style genres that can make highly polished content fall flat, ironically, as though the makers were trying too hard.

That said, there are still plenty of youth without easy access to the forms of high-speed connectivity and mobility that support content creation anytime and anywhere. Even media designed to look quick-and-dirty can require sophisticated staging and editing that can be done only with lots of practice and mentorship. Especially in schools with high concentrations of families living in poverty, we see too many examples of cases in which the equipment might be there but there is no curriculum that uses those resources to support higher-order thinking, critical engagement, and opportunities to apply lessons to novel situations—a cluster of abilities that S. Craig Watkins (2010, 2012) has labeled *critical design literacy*.

The specific demands associated with content-world creation as a tactic of participatory politics up the ante further. Young people need to know how to translate complex and nuanced issues relevant to public spheres into narratives with characters, plots, moods, and scenes that don't just tell a story but also invite engagement. The richest content worlds within participatory politics don't exist only online. As I've noted a few times and will revisit at the end, posting a video is not enough. You must create content that both fuels meaningful face-to-face interaction and builds momentum through online engagement (the Occupy movement stands out as especially effective in this sense, as explored by YPP's Harvard-based team).

To build mastery in these activities among youth, educators might give young people the challenge of reinventing the public service announcement, screening a series of corny examples as well as innovative social messaging experiments, and then invite the students to plan and pitch their own cross-platform campaigns on topics of their choosing. Critical curating abilities can also be a powerful first step onto a pathway that leads toward content-world creation. Young people might find media that are already in circulation and repost that material with additional writing and images in ways that advance a larger debate and invite further engagement.

The third tactic of participatory politics, foraging for information, consists of finding, sharing, and interpreting data available through proliferating social media and public archives to advance understanding and justice. To glean and package actionable information from a dizzying array of minimally vetted sources, young people need active support systems that help them tap and mine meaningful insights from complex data sets, including some that are walled off from easy public access, and to deliver these insights to target audiences.

Some of the academic subjects seen by many youth as deadly, based on how they're typically taught in school (e.g., math and statistics), can suddenly start to feel vital to them. To raise awareness and instigate action on issues they care deeply about, young people will increasingly be called upon to "show me the data," in rigorous and provocative displays. As Lindsay Grant (2012) has argued, though, "datafication" is freeing only to the extent that it enables young people to keep asking new and worthwhile questions. Debates about interpretation can be as productive and mind shifting as the conclusions that make their way into public spheres, and we need curriculum models that foster that spirit of iterative analysis.

Beyond the tasks of identifying and negotiating access to information troves, building the skills to fact-check and track patterns, and triangulating contradictory information, young people deploying the tactic of information foraging are also in the business of data *representation*. Here's where we start to see the exciting possible melding of technical and creative subjects (admittedly already a specious dichotomy). The ability to design a compelling infographic that people are inspired to share with friends, or to cleverly transform environmental pollution data into a musical soundtrack, emerges not as a neutral design assignment but as the possible fulfillment of a civic mandate.

The data young people need access to in order to move their issues forward will often be proprietary. An essential component of any curricular approach that supports this tactic will be mechanisms for youth to understand digital rights and advocate collectively for the transparency of the platforms and data sets from which the public has the most to gain and learn.

In coding up, the fourth tactic in our series, young people program tools and platforms that advance the public good. To do so, they need not only the concrete skills of computer programming but also the capacity for some specific forms of collective intelligence. Cathy Davidson (2012a, 2012b) has made the case that a fourth R should be added to the standard required literacy lineup of reading, 'riting, and 'rithmetic: 'rithms, as in algorithms, which she sees as the basis of computational thinking, coding, and webcraft.

Extending the technical definition of *algorithm* into a metaphor for the kind of thinking that enables young people to develop digital tools and platforms and not just use the existing ones, Davidson says literacy in this realm can't be postponed and reserved for college-bound kids. "What could be more relevant," she asks, "to the always-on student of today than to learn how to make apps and programs and films and journalism and multimedia productions and art for the mobile devices that, we know, are ubiquitous in the United States?" (2012b, n.p.).

Free tools like MIT's App Inventor and Mozilla's Hackasaurus and Popcorn Maker, as well as DIY education sites like Code Academy, are powerful resources that can introduce novices to code. But we need better systems and incentives to draw allies with engineering expertise into mentoring relationships with youth, and we need a curriculum that supports the production of civic software. Algorithms, after all, can do both harm and good.

I would add a second kind of literacy to this discussion of coding up. Young people need practice thinking constellationally as well as algorithmically. I'm borrowing a term from Teju Cole (2012), who used the term *constellational thinking* to conclude his critique, which I've already cited, of *Kony 2012*, which he published in the *Atlantic* shortly after that film caught fire (see also Rheingold 2012). Especially when political activity interferes in the lives of others, constellational thinking means always and only acting "with awareness of what else is involved" (Cole 2012).

Privilege and distance too often block constellational thinking by allowing us to impose solutions that ignore how even our best intentions can hurt people and that obscure how some of us benefit from the way things are. Coding can feel politically neutral, and algorithms can appear to offer tidy formulas for right answers (Wilson 2012). But if there's one thing we learn from the logic of programming, it's that everything is connected to everything else. One apparently tiny change can finally make the whole thing work, or it can screw the whole thing up. When young people aim to promote democracy, equity, and freedom, they deserve to be held to a standard that pushes them not just to fiddle with product design but also to interrogate the constellation of experiences their technology solutions can both create and trample.

The final tactic of participatory politics, hide and seek, involves engaging in civic activities that only selectively surface into public awareness. To express political speech while staying safe and managing privacy, young people need to understand what it takes to maintain diverse digital identities across networks governed by distinct and often nontransparent protocols for connection, encryption, and discovery. These protocols can, of course, obliterate the best-laid boundaries and juxtapose young people's various civic commitments with one another on a list of search results.

I have had direct experience with this dilemma, especially with young people who've been publishing revealing content about themselves and their politics starting at an early age. Every once in awhile, a colleague or I will get a call at Youth Radio from a graduate of the program requesting that we "unpublish" a commentary on a sensitive topic that aired years ago, sometimes because the author's positions have changed, sometimes out of fear of professional or social fallout, and sometimes for reasons I don't fully understand. The beauty of radio as a space for youth learning used to be that the story "evaporated into the ether," as my boss used to say, after the broadcast. Of course, now the post associated with the story persists forever, permanently attaching young people to their own teenage sentiments and to the comment streams their stories sparked.

In my view, among the best ways to support literacy development in this area is to learn from the young people who have the most at stake. The YPP research team headed by Henry Jenkins and Mimi Ito held a symposium in 2011 called "DREAMing Out Loud," which brought immigrant youth activists together to discuss their work at the intersection of digital media, art, and social justice.

More convenings along these lines, where youth organizers and media producers with a range of ideologies can share their experiments in participatory politics, are needed. Through these kinds of gatherings, both in real life and online, we can start to learn from young people's own best practices, as well as from their mistakes, in deciding what to expose and how to protect information that could damage lives and movements for justice.

We can also build organizations that expressly and strategically support these best practices, making them accessible to a greater number of youth.

Adult allies need to be prepared to support young people as they figure out where to draw their lines, personally and collectively, as they occupy sites where surveillance is present but not always obvious and where hypervisibility sometimes offers protection and at other times poses its own dangers. These allies also need to be willing to provide support grounded in seasoned ethics and to absorb at least some of the risks faced by young people who take a stand without a lot of institutional protection (Gardner 2013).

In closing this discussion of literacies associated with participatory politics, I want to be careful not to create three false impressions. First, we tend to associate literacies with skill sets taught by adults in formal educational institutions. As evident from the above discussion, young people themselves often mentor one another in the kinds of habits and practices that support effective forms of digital civics. Although I would argue that participatory politics are often strongest when young people and adults are in them together, I do not want to imply that adults are necessarily in the positions of authority to teach what young people need to know.

Second, my hesitation in even using the term *literacy* is that too often, lists like these are seen as exhaustive and, even more pernicious, are immediately used to rank and rate students on their relative levels of knowledge acquisition, highlighting deficiencies in those with fewer opportunities to learn. That is the last thing we need. Literacies are not skill sets possessed by individuals but practices we can cultivate within learning environments where young people are doing some of their most robust work advancing understanding and justice in public spheres.

Third, literacies imply goodness. Throughout this report, I have tried to highlight the potential of each tactic and also point to ways in which efforts to have a voice and exert influence in public spheres can backfire, making things worse. Building literacy in every case means learning how to carry out some of the most promising tactics within participatory politics and knowing how to regroup when one's efforts derail.

Mind the Risks

Now I will address a series of concerns that merit serious attention as we work to encourage the strengths and minimize the risks of next-generation civic engagement. Digital tools remove some of the barriers to civic participation, but they also eliminate some of the safeguards that have traditionally been in place to mitigate harm, and they can invite their own problems as well.

Simplification

Digital media conventions for production and circulation can compel citizens to sacrifice important nuance in the messages that drive movements. Brevity, such as the 140-character limit of Twitter or the assumption that videos won't "go viral" if they're more than a couple of minutes long, is often blamed for the dumbing down of civic discourse. Yet time and again we see politically trenchant Twitter feeds that manage to spark profound debate through short bursts of expression. In contrast, hour-long videos can be dismayingly and dangerously simplistic.

The underlying problem is not the inherent limits of any given format or genre. It's the belief that for a message to spread, it has to lack complexity or internal contradiction. It's true that any media product in isolation will never capture all there is to say, acknowledge every caveat, or consider every possible point of view. That's why we're calling for studies of participatory politics that account for bodies of work over time, for media that are both spreadable and drillable (Jenkins, Ford, and Green 2013) and for actions that don't reduce enduring conflicts with deep roots and far-flung implications to battles between good guys and bad guys.

The tools and tactics of participatory politics are, in fact, uniquely set up to reveal the hidden harm of what looks virtuous and the logic that can hide underneath something too easily dismissed as all wrong. That said, with more and more movements targeting change at the level of discourse, we run the risk of pursuing simple attention as the ultimate political currency, sometimes forgoing or at least postponing efforts to change something more concrete, like a law or a policy (Zuckerman 2012a).

Sensationalization

The pressure to simplify often triggers an urge to sensationalize: Let's find the most extreme, grotesque, and riveting manifestation of whatever civic issue is motivating our politics and heighten that story through digital media production and dissemination. This phenomenon is as old as every form of media itself. What's new is that young people are increasingly the ones creating the news, so they need to be aware of the ways in which their own productions can reify these familiar patterns.

Sensational stories can make for great media, but they distort the truth. In creating compelling content worlds, we can't lose sight of scale. How representative is this story? Who benefits from this telling? What will it mean for those profiled in any given account to be presented in this light? Just because young people are among those who have suffered the most from media sensationalization doesn't mean they're immune to the instinct to tell the most attention-grabbing story. What can get lost are the efforts to dislodge more mundane realities that reinforce the status quo, as well as the less glamorous grind of pursuing legislative and policy change.

Slippage

For civically engaged youth who follow a range of political-thought leaders and movements, any given day's social media feed can play like a surreal simulcasting of disparate struggles—local, national, and global. The challenge that comes with this weird juxtaposition of dispatches is that it becomes easy to assume that the dynamics governing the causes you actually know a lot about are universally relevant. Conditions that differentiate struggles and call for specific forms of organizing can slip out of focus.

With many of the most politically charged issues in recent memory, there has been a curious pattern of community members claiming a kind of one-click solidarity that moves beyond the message "I get you" or "I'm with you" all the way to "I am you" or even "We are all you." This sentiment has echoed through the "We are the 99 percent" discourse of the Occupy movement; photos of white people in hoodies to signal their support for Trayvon Martin, the black teen who was killed by a neighborhood watchman in Florida; "We are all Khaled Said" status updates on the case of a young Egyptian man beaten to

death by policemen during the Arab Spring; and even the contro-
versy around a blog post asserting "I am Adam Lanza's mother"
in the wake of the Sandy Hook Elementary School shootings in
Newtown, Connecticut, written by a woman who believed her
own mentally ill son to be capable of chilling violence.

It can be advantageous and enriching to focus on common-
alities that unite our struggles and to insist on the possibility of
empathy across disparate identities and experiences. But there
is also the risk here that conventions in digital shorthand gloss
over such inequalities as class, race, geography, and disability,
which must be seen for what they are if participatory politics are
to advance freedom and justice in public spheres.

Unsustainability

In participatory politics, as in many other things in life, getting
started is often a lot easier than keeping something going. In the
wake of a specific crisis, like Hurricane Sandy in New York and
New Jersey, we can see an intense and hopeful flurry of public
response that dwindles quickly as people return to their daily
lives. Even with persistent community problems—like the inac-
cessibility of high-quality, affordable, fresh food in low-income
communities—sometimes developers or funders will become
energized to create prototypes for solutions but fall short of
resources, both human and monetary, to build the level of
enduring engagement necessary to make those interventions
take root and grow.

The challenge is to manage expectations from the outset of
an undertaking and to set a realistic scope and plan for any given
effort's "end-of-life decisions" (as my Youth Radio colleague
dramatically calls the need for a clear handoff and postlaunch

strategy for every app created in our lab). Otherwise, under-resourced communities already subjected to inadequate and inconsistent public support find themselves dealing with the aftermath of empty platforms, glitchy sites, stalled efforts, and broken commitments.

Saviorism

When distance shrinks and young people are exposed to far-away struggles without sufficient context, saviorism can set in. Through well-meaning civic engagements, those who are already relatively empowered to "do good" and "make a difference" can lead with their own needs, reproducing privilege and worse.

That much has been established, I hope, throughout this report, so I won't repeat the points here. But I would just add that this dynamic isn't an issue only in cases of global activism like the events surrounding *Kony 2012*. Consider another 2012 video sensation, *Caine's Arcade*. This tells the story of a nine-year-old boy who cobbled together a magnificent arcade out of cardboard, stuffed animals, and plastic toys, all of it held together with a dazzling hodgepodge of pipe cleaners, pushpins, colored yarn, and see-through duct tape.

Caine operated the arcade out of his dad's East Los Angeles auto-body shop. A local filmmaker happened to come into the shop, saw Caine's installation, and was totally inspired. (You'd pretty much need to have a heart of stone not to be.) So he decided to make a film about it. The filmmaker got the idea to surprise Caine with a huge crowd of visitors, so in cahoots with Caine's dad, he arranged for a humongous crowd of Angelenos to descend on the auto-body shop while Caine was out for lunch.

In the video, you watch the boy arrive on the scene strapped into the seat of a car, giggling and beaming when he sees the cheering crowd. The filmmaker greets Caine with a microphone and says to the throng, "Welcome to Caine's arcade, man." It's a genius, chills-inducing cinematic moment and might very well be a main reason the video became such a sensation, launching a scholarship campaign that is likely to make a real difference in Caine's life.

In addition to the achievement itself and the filmmaker's commitment and brilliance, the event is also a moment when Caine's position shifts from host to interviewee, from a maker who masterminded an elaborate invention to a kid arriving at his own surprise party. (Actually, another way to look at it is that Caine got to be all four of these things at once.) The larger point is that in our efforts to join forces with young people at their most creative and powerful, it's probably worthwhile to watch for moments like these so that we're aware of the new dynamics that media attention and adult involvement can set in motion when young people's voices are heard in public spheres in a big way.

Concluding Thoughts

In this report, I have drawn from the Youth and Participatory Politics Research Network and other sources to identify a set of emerging tactics young people are using to engage with and remake public spheres, often deploying digital and social media tools in intriguing ways. I have linked those tactics to a series of literacies that young people will increasingly rely on as they exercise civic agency. I have also highlighted some concerns related to participatory politics—vulnerabilities in the model that can cause even well-intentioned efforts to do inadvertent harm.

Table 1 shows how the various dimensions of participatory politics can work together.

The configuration in table 1 is just one snapshot of how the features, tactics, literacies, and risks of participatory politics can correspond to one another. For example, in the first row I've posited a scenario in which young people create content worlds primarily to circulate civic media; to do so they utilize an ability to collaborate in making stories that engage audiences in strategic ways, and they need to be very intentional about not sacrificing complexity and understanding for the sake of dissemination.

Table 1

Feature	Tactic	Literacy	Risk
Circulation	Create content worlds	How to participate in transmedia storytelling that inspires sharing and action.	Simplification: Nuance in the message can get lost, and dissemination can get out of control.
Dialogue and feedback	Hide and seek	How to structure collective interaction in ways that selectively disclose personal and other forms of highly sensitive information in order to generate substantive conversation within ideologically diverse communities.	Sensationalization: Stories with the most epic and still palatable themes capture public attention. Everyday experiences of ordinary lives shadowed by fear and constrained by unfair policies don't rise to that level of attention and therefore fail to factor in debates and decisions.
Production	Code up	How to design and develop platforms that invite and constrain modes of engagement toward desired ends, balancing openness and the aim to engineer specific forms of user response.	Unsustainability: Civic software projects can require ongoing iteration over extended periods—a never-ending cycle that is hard to maintain with limited resources; even the most promising abandoned experiments can do more harm than good.
Investigation	Forage for information	How to glean insights from dense databases that aren't always set up to invite access or scrutiny, and then how to represent the findings in accurate and compelling ways.	Saviorism: The zeal for "making a difference" on behalf of others perceived as vulnerable can lead young people (and the rest of us) with an incomplete understanding of a complex situation to misconstrue information and release it in ways that can be dangerous to self and others.
Mobilization	Pivot your public	How to support one's peers and others in ways that invite reciprocity when the time is right to enlist one's network to take action on a specific issue or cause.	Slippage: The desire to utilize shared interests to mobilize peers on behalf of a deeply felt cause can blind us to consequential differences, gaps in our own understanding, and limits to our own solidarity.

Although the tactic of creating content worlds has been paired with circulation in the table, it can relate to the other features as well. It is extremely relevant to production, it (one hopes) invites substantive dialogue and feedback, it can require investigation to get the story straight, and it can be a part of a larger campaign designed to mobilize a particular form of collective action.

Likewise, young people seeking to create content worlds for civic ends would need to look out for all the risks associated with participatory politics, not just simplification. Content worlds can feed sensationalization; they can ultimately be unsustainable and thus set up the participants for disappointment, resentment, cynicism, and missed opportunity; they can reveal a kind of saviorism that denies agency to those with direct knowledge and the most to lose; and they can invite slippage to the extent that participants eager to connect with the widest possible audience sometimes obscure the specificity of particular struggles.

The fluidity in how these features, tactics, literacies, and risks connect helps explain the potential power of participatory politics at its most fully realized. There are multiple possible points of entry that tap a range of capacities within individuals and communities to create positive change. The flexibility within the model also highlights the challenges involved: all the other risks are always hovering and require vigilance if the people involved are, at minimum, to do no harm.

I'd like to end with some final ruminations on the underlying social dynamics that participatory politics help bring to light. We're seeing a strengthening of ties between politics and everyday creativity, lowered barriers to entering civic efforts, greater recognition for young people as producers of media and culture, and evidence that they're utilizing traditional organizations in new ways, sometimes bypassing or installing new gatekeepers. We're seeing a shift in how information and individuals accrue trust, credibility, and influence, not so much through official certification but more and more by way of association with valued networks and searchable track records of activity.

A defining feature of participatory politics is its center of gravity in peer relationships. Young people can find civic resources within their own communities, and not all their efforts necessarily aim at the usual targets. With widespread distrust of the formal institutions of government and conventional mechanisms for creating change, young people are experimenting with bottom-up tactics to challenge the social order. That said, face-to-face, sustained adult mentoring is still key in young people's stories of political becoming. Allies who have injected equal doses of time, expertise, and humility into collaborative civic work with youth are indispensable (Gardner et al. 2012). We live together, after all, in these public spheres.

Participatory politics at its most influential often has a "transline" quality—not online or offline, but both. The convergence of these spheres of experience reflects a trend that's accelerating, according to Wendy Hui Kyong Chun (forthcoming).[1] One way to understand the significance of this transline quality for participatory politics is to recognize face-to-face interaction in shared physical space as its own kind of medium, not unlike video, audio, or text messaging.

Like any of these forms, the medium of live, to borrow a phrase from writer Douglas McGray, has its own special affordances and limits. It invites characteristic behaviors and inhibits others. Thus, among the most striking developments reported here—something that says a lot about how the world has changed—is the possibility that framing our questions about youth civics in ways that isolate the digital difference might, in the not-too-distant future, start to seem less and less like a good idea. The affordances of new media are key, but they are also inexorably enmeshed with the offline practice of politics. We therefore seek to understand how young people are producing civics today, seizing every tool, platform, and structure they can find or else cocreate.

Notes

Five Tactics of Participatory Politics

Please note that the quotes in this chapter without retrievable sources are either from personal communications or from research still ongoing and as yet unpublished.

1. Invisible Children research is supported by the John D. and Catherine T. MacArthur Foundation and the Spencer Foundation, through funding to Henry Jenkins and his colleagues at the University of Southern California.

2. See the Invisible Children's vimeo site, at http://vimeo.com/invisible.

3. See Apps for Communities, at http://appsforcommuni ties.challenge .gov.

4. See the Human Face of Big Data's website, at http://humanfaceofbig data.com/about.

5. The Mobile Action Lab is funded in part by the MacArthur Foundation and the National Science Foundation.

6. Asha Richardson and I cofounded the Lab, and since 2012 it has been run by Kurt Collins.

Concluding Thoughts

1. The blending of online and offline might appear to make the Web safer and less nasty by diminishing anonymity, but Chun says that it's often motivated by monetization (advertisers pay premiums when they know whose eyeballs they're attracting) and that it sometimes fosters greater cruelty when users know whom they're dealing with and can track adversaries beyond the digital realm.

References

Allen, D. 2010. "Anonymous: On Silence and the Public Sphere." In *Speech and Silence in American Law*, ed. A. Sarat, pp. 106–133. Cambridge: Cambridge University Press.

Allen, D. 2012. *Toward Participatory Democracy*. Boston: Boston Review.

Allen, D., and J. Light, eds. Forthcoming. *Transforming Citizens: Youth, Media, and Political Participation*.

Barron, B., K. Gomez, N. Pinkard, and C. K. Martin, eds. Forthcoming. *The Digital Youth Network: Cultivating New Media Citizenship in Urban Communities*. Cambridge, MA: MIT Press.

Bourdieu, P. 1977. *Outline of a Theory of Practice*. Cambridge, UK: Cambridge University Press.

boyd, d. 2010. "Privacy and Publicity in the Context of Big Data." Paper presented at WWW, Raleigh, NC, April 29.

boyd, d., H. Thakor, M. Casteel, and R. Johnson. 2011. *Human Trafficking and Technology: A Framework for Understanding the Role of Technology in the Commercial Sexual Exploitation of Children in the U.S.* http://research.microsoft.com/en-us/collaboration/focus/education/htframework-2011.pdf.

Chaiklin, S., and J. Lave, eds. 1996. *Understanding Practice: Perspectives on Activity and Context*. Cambridge, UK: Cambridge University Press.

Chen, M. 2011. "Activists Enter Detention and Emerge Inspired." Culture Strike, http://culturestrike.net/activists-enter-detention-and-emerge-inspired.

Chun, W. Forthcoming. "The Dangers of Transparent Friends: Crossing the Public and Intimate Spheres." In *Transforming Citizens*, ed. D. Allen and J. Light. Cambridge, MA: MIT Press.

Cohen, C., J. Kahne, B. Bowyer, E. Middaugh, and J. Rogowski. 2012. *Participatory Politics: New Media and Youth Political Action*. Digital Media and Learning Central, http://ypp.dmlcentral.net/sites/all/files/publications/YPP_Survey_Report_FULL.pdf.

Cole, T. 2012. "The White-Savior Industrial Complex." *Atlantic*, March, http://www.theatlantic.com/international/archive/2012/03/the-white-savior-industrial-complex/254843.

Davidson, C. 2012a. *Now You See It: How Technology and Brain Science Will Transform Schools and Business for the 21st Century*. New York: Penguin.

Davidson, C. 2012b. "A Fourth R for 21st Century Learning." *Washington Post*, January 2.

DeVoss, D. N., E. Eidman-Aadahl, and T. Hicks. 2010. *Because Digital Writing Matters: Improving Student Writing in Online and Multimedia Environments*. San Francisco: Jossey-Bass.

Dougherty, D. 2012. "Maker Spaces in Education and DARPA." Makezine. http://makezine.com/2012/04/04/makerspaces-in-education-and-darpa.

Earl, J., and K. Kimport. 2011. *Digitally Enabled Social Change: Activism in the Internet Age*. Cambridge, MA: MIT Press.

Gardner, H. 2013. "Re-Establishing the Commons for the Common Good." *Daedalus* 142, no. 2:199–208.

Gardner, H., C. James, D. Knight, and M. Rundle. 2012. "Diverse Pathways of Participatory Politics." Paper presented at Youth and Participatory Politics Research Network, Boston, April.

Glassman, J., and J. Cohen. 2008. "Special Briefing to Announce the Alliance of Youth Movement." U.S. Department of State Archive, http://2001-2009.state.gov/r/us/2008/112310.htm.

Grant, L. 2012. "Datafication: How the Lens of Data Changes How We See Ourselves." DML Central, http://dmlcentral.net/blog/lyndsay-grant/datafication-how-lens-data-changes-how-we-see-ourselves.

Grossman, J., C. Chan, S. Schwartz, and J. Rhodes. 2012. "The Test of Time in School-Based Mentoring: The Role of Relationship Duration and Re-Matching on Academic Outcomes." *American Journal of Community Psychology* 49:43–54.

Heath, S. B., J. Flood, and D. Lapp. 2008. *Lawrence Handbook for Literacy Educators: Research in the Visual and Communicative Arts, 2.* New York: Erlbaum.

Hing, J. 2011. "Alabama DREAMers Speak from Detention: ICE Is 'Rogue Agency.'" Colorlines, http://colorlines.com/archives/2011/11/dreamers _in_detention_expose_obamas_deportation_lies.html.

Howard, P. 2010. *The Digital Origins of Dictatorship and Democracy: Information Technology and Political Islam.* Oxford, UK: Oxford University Press.

Invisible Children Inc. 2012. Independent Auditors' Report. http://invisiblechildrencom.files.wordpress.com/2012/12/annotated_finan cialsfy2012.pdf.

Ito, M. 2009. *Hanging Out, Messing Around, and Geeking Out: Kids Living and Learning with New Media.* Cambridge, MA: MIT Press.

James, C. 2012. Personal communication.

Jenkins, H. 2008. *Convergence Culture: Where Old and New Media Collide.* New York: New York University Press.

Jenkins, H. 2012. "DREAMing Out Loud! Youth Activists Spoke about Their Fight for Education, Immigrant Rights and Justice through Media and Art." http://henryjenkins.org/2012/01/dreaming_out_loud_youth _activi.html.

Jenkins, H., S. Ford, and J. Green. 2013. *Spreadable Media: Creating Value and Meaning in a Networked Culture*. New York: New York University Press.

Jenkins, H., R. Purushotma, K. Clinton, M. Weigel, and A. J. Robison. 2009. *Confronting the Challenges of Participatory Culture: Media Education for the 21st Century*. Chicago: John D. and Catherine T. MacArthur Foundation.

Jones, J. 2011. "The Network Society after Web 2.0: What Students Can Learn from Occupy Wall Street." Digital Media and Learning Central, http://dmlcentral.net/blog/john-jones/network-society-after-web-20-what-students-can-learn-occupy-wall-street.

Kahne, J., E. Middaugh, and D. Allen. Forthcoming. "Youth, New Media and the Rise of Participatory Politics." In *Transforming Citizens*, ed. D. Allen and J. Light.

Kligler-Vilenchik, N. Forthcoming. Fan Communities Case Study.

Kligler-Vilenchik, N., and S. Shresthova. 2012. *Learning through Practice: Participatory Culture Civics*. Digital Media and Learning Central, http://dmlhub.net/publications/learning-through-practice-participatory-culture-practices.

Kron, J. 2012. "Mission from God: The Upstart Christian Sect Driving Invisible Children and Changing Africa." *Atlantic*, April 10.

Lenhart, A. 2012. *Cell Phone Ownership*. Pew Internet and American Life Project, http://pewinternet.org/Reports/2012/Teens-and-smartphones/Cell-phone-ownership/Smartphones.aspx.

Madrigal, A. 2012. "Dark Social: We Have the Whole History of the Web Wrong." http://www.theatlantic.com/technology/archive/2012/10/dark-social-we-have-the-whole-history-of-the-web-wrong/263523.

Maira, S. 2011. *Missing: Youth, Citizenship, and Empire after 9/11*. Durham, NC: Duke University Press.

Middaugh, E. 2012. *Service and Activism in the Digital Age: Supporting Youth Engagement in the Public Life*. DML Central, http://www.civicsurvey.org/Service_Activism_Digital_Age.pdf.

Niemeyer, G., A. Garcia, and R. Naima. 2009. "BlackCloud: Patterns Towards da Future." *Proceedings of the Seventeenth ACM International Conference on Multimedia*, 1073–1082. Beijing: Association for Computing Machinery.

Ortner, S. 1984. "Theory in Anthropology since the sixties." *Comparative Studies in Society and History* 26, no. 1:126–166.

Phillips, W. 2011. "LOLing at Tragedy: Facebook Trolls, Memorial Pages, and Resistance to Grief Online." *First Monday* 16, no. 12. http://first monday.org/ojs/index.php/fm/article/view/3168.

Pratt, M. L. 1991. "Arts of the Contact Zone." *Profession*. http://www.jstor.org/discover/10.2307/25595469?uid=3739560&uid=2&uid=4&uid=3739256&sid=21102544142027.

Rheingold, H. 2012. *Net Smart: How to Thrive Online*. Cambridge, MA: MIT Press.

Schrock, A. 2011. "Hackers, Makers and Teachers: A Hackerspace Primer." http://andrewrschrock.wordpress.com/2011/07/27/hackers -makers-and-teachers-a-hackerspace-primer-part-1-of-2.

Sefton-Green, J. 2013. "Mapping Digital Makers: A Review Exploring Everyday Creativity, Learning Lives and the Digital." Nominet Trust State of the Art Reviews. http://www.nominettrust.org.uk/sites/default/ files/NT%20SoA%206%20-%20Mapping%20digital%20makers.pdf.

Soep, E. 2005. "Making Hard-Core Masculinity: Teenage Boys Playing House." In *Youthscapes: The Popular, the National, the Global*, ed. S. Maira and E. Soep, 173–191. Philadelphia: University of Pennsylvania Press.

Soep, E. 2011. "All the World's an Album: Youth Media as Strategic Embedding." In *International Perspectives on Youth Made Media*, ed. J. Fisherkeller, 246–262. New York: Peter Lang.

Soep, E. 2012. The Digital Afterlife of Youth Made Media. *Comunicar* 38, no. 19:93–100.

Soep, E., and V. Chávez. 2010. *Drop That Knowledge: Youth Radio Stories*. Berkeley: University of California Press.

Street, B., ed. 2001. *Literacy and Development: Ethnographic Perspectives.* New York: Routledge.

Swartz, L. 2012. *Invisible Children.* Civic Paths, http://civicpaths.uscan nenberg.org/wp-content/uploads/2012/03/Swartz_InvisibleChildren _WorkingPaper.pdf.

Thompson, L. G. 2012. *The Cost of Engagement: Politics and Participatory Practices in the U.S. Liberty Movement.* DML Central, http://ypp.dmlcen tral.net/sites/all/files/publications/The_Cost_of_Engagement-Working _Paper-MAPP_12.10.12.pdf.

Varenne, H., and R. McDermott. 1999. *Successful Failure: The School America Builds.* Boulder, CO: Westview Press.

Watkins, S. C. 2010. *The Young and the Digital: What the Migration to Social Network Sites, Games, and Anytime, Anywhere Media Means for Our Future.* Boston: Beacon Press.

Watkins, S. C. 2012. "From Theory to Design: Exploring the Power and Potential of 'Connected Learning,' Part Two." The Young and the Digital, http://theyoungandthedigital.com/2012/10/09/from-theory-to -design-exploring-the-power-potential-of-connected-learning-part-2.

Weinstein, E. 2013. "Beyond Kim Kardashian on the Middle East: Patterns of Social Engagement among Civically-Oriented Youth." Good Project, http://www.thegoodproject.org/beyond-kim-kardashian-on -the-middle-east-patterns-of-social-engagement-among-civically -oriented-youth.

Wilson, G. 2012. "On Algorithmic Thinking." Third Bit, http://third-bit .com/blog/archives/4426.html.

Zimmerman, A. 2012. *Documenting Dreams: New Media, Undocumented Youth and the Immigrant Rights Movement.* DML Central, http://dmlcen tral.net/resources/5061.

Zuckerman, E. 2008. "The Cute Cat Theory Talk at ETech." http://www .ethanzuckerman.com/blog/2008/03/08/the-cute-cat-theory-talk-at -etech.

Zuckerman, E. 2012a. "Attention Activism and Advocacy in the Digital Age." Connected Learning, http://connectedlearning.tv/ethan-zucker man-attention-activism-and-advocacy-digital-age.

Zuckerman, E. 2012b. "Unpacking Kony 2012." My Heart's in Accra, http://www.ethanzuckerman.com/blog/2012/03/08/unpacking -kony-2012.

Zuckerman, E. Forthcoming. "Cute Cats to the Rescue? Participatory Media and Political Expression." In *Transforming Citizens*, ed. D. Allen and J. Light.

The John D. and Catherine T. MacArthur Foundation Reports on Digital Media and Learning

Peer Participation and Software: What Mozilla Has to Teach Government by David R. Booth

Kids and Credibility: An Empirical Examination of Youth, Digital Media Use, and Information Credibility by Andrew J. Flanagin and Miriam Metzger with Ethan Hartsell, Alex Markov, Ryan Medders, Rebekah Pure, and Elisia Choi

The Future of Thinking: Learning Institutions in a Digital Age by Cathy N. Davidson and David Theo Goldberg with the assistance of Zoë Marie Jones

New Digital Media and Learning as an Emerging Area and "Worked Examples" as One Way Forward by James Paul Gee

Digital Media and Technology in Afterschool Programs, Libraries, and Museums by Becky Herr-Stephenson, Diana Rhoten, Dan Perkel, and Christo Sims with contributions from Anne Balsamo, Maura Klosterman, and Susana Smith Bautista

Young People, Ethics, and the New Digital Media: A Synthesis from the GoodPlay Project by Carrie James with Katie Davis, Andrea Flores, John M. Francis, Lindsay Pettingill, Margaret Rundle, and Howard Gardner

Confronting the Challenges of Participatory Culture: Media Education for the 21st Century by Henry Jenkins (P.I.) with Ravi Purushotma, Margaret Weigel, Katie Clinton, and Alice J. Robison

The Civic Potential of Video Games by Joseph Kahne, Ellen Middaugh, and Chris Evans

Quest to Learn: Developing the School for Digital Kids by Katie Salen, Robert Torres, Loretta Wolozin, Rebecca Rufo-Tepper, and Arana Shapiro

Measuring What Matters Most: Choice-Based Assessments for the Digital Age by Daniel L. Schwartz and Dylan Arena

Learning at Not-School? A Review of Study, Theory and Advocacy for Education in Non-Formal Settings by Julian Sefton-Green

Measuring and Supporting Learning in Games: Stealth Assessment by Valerie Shute and Matthew Ventura

Participatory Politics: Next-Generation Tactics to Remake Public Spheres by Elisabeth Soep *The Future of the Curriculum: School Knowledge in the Digital Age* by Ben Williamson

For more information, see http://mitpress.mit.edu/books/series/john-d-and-catherine-t-macarthur-foundation-reports-digital-media-and-learning.